Neoliberal Environments

Surprisingly, very little work has systematically explored the application of neoliberal policies to environmental governance and environmental change. This volume explores this nexus between nature, markets, deregulation and valuation, using theoretically sharp and empirically rich real-world case studies and analyses of actually existing policy from around the world and across a range of resources. In short, it answers the questions: does neoliberalizing nature work and what work does it do? More specifically, this volume provides answers to a series of urgent questions about the effects of neoliberal policies on environmental governance and quality. What are the implications of privatizing public water utilities in terms of equity in service provision, resource conservation and water quality? Do free trade agreements erode the sovereignty of nations and citizens to regulate environmental pollution, and is this power being transferred to corporations? What does the evidence show about the relationship between that marketization and privatization of nature and conservation objectives?

Neoliberal Environments productively engages with all of these questions and more. At the same time, the diverse case studies collectively and decisively challenge the orthodoxies of neoliberal reforms, documenting that the results of such reforms have fallen far short of their ambitions.

Nik Heynen is Assistant Professor of Geography at the University of Georgia.

James McCarthy is Associate Professor of Geography at Penn State University.

Scott Prudham is Associate Professor in the Department of Geography and the Centre for Environment at the University of Toronto.

Paul Robbins is Professor of Geography at the University of Arizona.

Neoliberal Environments

False promises and unnatural consequences

Edited by
Nik Heynen, James McCarthy,
Scott Prudham, and Paul Robbins

LONDON AND NEW YORK

First published 2007
by Routledge
4 Park Square, Milton Park, Abingdon, Oxon OX14 4RN
605 Third Avenue, New York, NY 10017

Routledge is an imprint of the Taylor & Francis Group, an informa business

© 2007 Nik Heynen, James McCarthy, Scott Prudham, and Paul Robbins
for editoral and selection; the contributors, their contributions

Typeset in Times by
Taylor & Francis Books

British Library Cataloguing in Publication Data
A catalogue record for this book is available from the British Library

Library of Congress Cataloging in Publication Data
A catalog record for the book has been requested

ISBN 978–0–415–77148–1 (hbk)
ISBN 978–0–415–77149–8 (pbk)
ISBN 978–0–203–94684–8 (ebk)

Contents

Figures

Contributors

Karen Bakker is an Associate Professor in the Department of Geography at the University of British Columbia. Her research interests include environmental policy, natural resource management, and water governance.

Nick Blomley is a Professor in the Department of Geography at Simon Fraser University. His main research interest is the interface of law and space.

Neil Brenner is a Professor in the Department of Sociology at New York University. His main research interests include urban sociology, urban political economy and urban theory.

Gavin Bridge is a Reader in Economic Geography in the School of Environment and Development at Manchester University. His research interests include environmental governance and the social regulation of resource access.

Noel Castree is a Professor in the School of Environment and Development at Manchester University. His principal research interests have been in capitalism-environment relationships and how best to theorize them.

David Correia is an Assistant Professor of Geography at the University of Maine, Farmington. His research interests revolve around the political ecology of forest and rangeland management in New Mexico.

Jim Glassman is an Associate Professor in the Department of Geography at the University of British Columbia. His research interests are currently focused on the political economy of development in Southeast Asia

Nik Heynen is an Assistant Professor in the Department of Geography at the University of Georgia. His research interests include urban political economy/ecology, social theory, and inequality and social movements.

Ryan Holifield is a PhD Candidate at the University of Minnesota. He is most directly working on issues of environmental justice.

Gail Hollander is an Assistant Professor in the Department of International Relations and Geography at Florida International University. Her research interests include agricultural, environmental, and economic geographies, particularly in Florida and the Caribbean.

Roger Keil is a Professor in the Faculty of Environmental Studies and Department of Political Science at York University, Canada. His main areas of interest and research are urban politics and governance and urban political ecology.

Wendy Larner is a Professor in the School of Geographical Sciences at the University of Bristol. Her research interests include globalization, governance and gender.

April Luginbuhl is a graduate student in the Department of Geography at the Ohio State University. Her research focuses broadly on nature/society interactions, particularly the social and property relations of regulating greenhouse gases.

Becky Mansfield is an Assistant Professor in the Department of Geography at the Ohio State University. Her research interests are on the political economy of the environment, with a special focus often on ocean fisheries.

James McCarthy is an Associate Professor in the Department of Geography at Penn State University. His research centers on political ecology and political economy, with particular interests in environmental politics, conflicts, and social movements.

Nancy Lee Peluso is a Professor in Environmental Science, Policy and Management at the University of California, Berkeley. Her interests include a special focus on the social processes that affect the management of land-based and coastal resources.

Harold A. Perkins is an Assistant Professor in the Department of Geography at Ohio State University. His research interests are in connections between urban political economy and urban political ecology.

Scott Prudham is an Associate Professor in the Department of Geography and the Centre for Environment at the University of Toronto. He is interested in the environmental problems that arise (political, economic, ecological, cultural) from commodity production.

Paul Robbins is an Associate Professor in the Department of Geography and Regional Development at the University of Arizona. His research centers on the relationships between individuals (homeowners, hunters, professional foresters), environmental actors (lawns, elk, mesquite trees), and the institutions that connect them.

Morgan M. Robertson is an Assistant Professor in the Department of Geography at the University of Kentucky. His research interests include political ecology, the commodification of nature, and wetland ecology and policy.

Dianne Rocheleau is a Professor in the Department of Geography at Clark University. Her interests include environment and development, political ecology, forestry, agriculture and landscape change, with an emphasis on the role of gender, class and "popular" vs. "formal" science in resource allocation and land use.

Kevin St. Martin is an Assistant Professor in the Department of Geography at Rutgers University. His research interests include economic geography, diverse economies, political ecology, and community and commons.

Juanita Sundberg is an Assistant Professor in the Department of Geography at the University of British Columbia. Her research interests focus on conservation and development; gendered landscapes, and environmental thought.

Erik Swyngedouw is a Professor in Geography in the School of Environment and Development at Manchester University. His research has mostly been on the political economy of governance, geographical scale and nature.

Nik Theodore is the Director of the Center for Urban Economic Development and an Assistant Professor in the Urban Planning and Policy Program at the University of Illinois, Chicago. His research interests are in economic development, labor markets, and urban policy.

Michael Watts is a Professor in the Department of Geography and the Director of the Institute of International Studies at the University of California, Berkeley. His research interests include political economy, political ecology, development, peasant societies, social and cultural theory, U.S. agriculture, Islam and social movements.

Wendy Wolford is an Associate Professor in the Department of Geography at the University of North Carolina at Chapel Hill. Her research interests include the political economy of development, social movements, land reform and agrarian studies.

Douglas Young is a PhD Candidate in the Faculty of Environmental Studies at York University, Canada. His research interests include urban governance, urban planning and urban ecology.

Acknowledgments

The editors would like to acknowledge all the authors, including the authors of the commentaries.

We would like to thank Andrew Mould, Jennifer Page at Routledge and Zoe Kruze, formerly at Routledge, for all of their assistance.

We would like to thank Andrew Leyshon, who was the *Geoforum* editor who worked with James and Scott on our special issue; and Karen Charman and Joel Kovel who are the managing editor and editor-in-chief respectively at *Capitalism, Nature, Socialism* and helped Paul and Nik complete our special issue. We would also like to thank Jamie Peck for his insight while we worked on this volume.

Finally, we would like to thank our students. Teaching and discussing environmental politics and political ecology in classrooms has been invaluable in sharpening our thinking on these topics.

Introduction

False promises

Nik Heynen, James McCarthy, Scott Prudham, and Paul Robbins

> Meanwhile, capitalism rewards efficiency and punishes waste. Profit-hungry com-
> panies found ingenious ways to reduce the natural resource inputs necessary to
> produce all kinds of goods, which in turn reduced environmental demands on
> the land and the amount of waste that flowed through smokestacks and water
> pipes. As we learned to do more and more with a given unit of resources, the
> waste involved (which manifests itself in the form of pollution) shrank ... Prop-
> erty rights – a necessary prerequisite for free market economies – also provide
> strong incentives to invest in resource health. Without them, no one cares about
> future returns because no one can be sure they'll be around to reap the gains.
> Property rights are also important means by which private desires for resource
> conservation and preservation can be realized. When the government, on the
> other hand, holds a monopoly on such decisions, minority preferences in devel-
> oping societies are overruled (see the old Soviet bloc for details) ... Capitalism
> can save more lives threatened by environmental pollution than all the environ-
> mental organizations combined.
>
> (Jerry Taylor, Director of Natural Resource Studies, Cato Institute, 2003)

> Q: Do you think social movements are now on the offensive?
> A: Yes, for me they moved to the offensive a long time ago. I always say that on
>   February 27, 1989, the fourth world war began in Venezuela. The third had
>   been the Cold War, and the fourth is the war against neoliberalism, which
>   began in Caracas on that day.
> Q: As the 6th WSF [World Social Forum] continues, have you seen a main issue
>   of debate emerge?
> A: All of the issues are central. It would be trite to refer to a dominant issue.
>   Without ecology, we would all die; without the fight against poverty, ethnic,
>   racial, classist discrimination, our societies would always be a confrontation
>   of oppressors and the oppressed. Without the debate on gender issues,
>   patriarchy would last forever, or would be replaced by maternalism, for
>   example. It is precisely the social movements that emerge because the states
>   and the parties thought these were issues that didn't merit their attention.
>
> (Luis Britto Garcia, Venezuelan author and activist, 2006)

Both these quotes in their own way point to something important, if not
radical, that has recently changed in environmental governance; this change

is neoliberalism. The quotes obviously offer fundamentally different views on the character of the phenomenon and the role of politics and political struggle in both advancing and resisting the relationships between democracy, markets, the environment and the state. Yet both Taylor and Garcia recognize the significance of so-called free market reforms, institutionalized and globalized (i.e. launched across international contexts and more widely adopted) over the course of recent decades. Moreover, each asserts the centrality of economy and ecology to governance, social change, and politics. We would agree.

In fact, while these quotes highlight and reinforce the controversial character of neoliberalism, the widely diverging viewpoints on what neoliberalism represents, and whether or not neoliberalism is viewed positively or negatively, they also convey something of the central objective of this collection: to critically examine the somewhat overlooked nexus between neoliberalism, on the one hand, and environmental governance, environmental change, and environmental politics, on the other. Our primary goal in this respect is analytical: we recognize that the dynamics and outcomes of the sea-change that these radical reforms represent are inevitably complex and not entirely predictable. In that context, we seek to add to this contentious discussion by bringing into more intimate contact a collection of recent, rigorous, theoretically informed case studies. Each of these addresses in detail a range of reforms in environmental governance and the political and ecological outcomes they produce, but together we believe they tell a larger story. We also present commentaries in response to groups of substantively and thematically related cases, therefore, in order to provide coherent and current perspectives on what has been learned on the neoliberal journey so far, to highlight what we have yet to understand and explore.

As will become more evident in the collection and in the text to follow, we are by no means neutral on the questions examined here: rather, we believe that logic and the weight of the empirical evidence available so far strongly suggest that the so-called "neoliberalization"[1] of environmental governance will produce predominantly environmentally undesirable and socially regressive political and economic outcomes. That said, the volume is quite intentionally aimed at moving past mere polemics in the debate over environmental neoliberalism and mapping out a picture of the world that has appeared in its wake.

Neoliberalism, geography, environment

In order to contextualize the case studies and commentaries to come, some discussion seems warranted regarding the focus of the collection, begging two questions: (1) what exactly do we mean by the terms "neoliberalism" and "neoliberalization"?; and (2) what exactly are the connections between neoliberalism and environment? As we argue below, and as we hope the cases help to reinforce, these are not distinct questions. Indeed, we believe that environmental change and environmental politics are in substantial measure constitutive of the brief history of neoliberalism in important and yet largely overlooked ways, not least in the ways that attempts to "stretch" and "deepen" (Lysandrou 2005) the reach of commodity circulation rely on the re-working of environmental governance and on entrenching the commodification of nature, and vice versa.

But first, neoliberalism. It is a big word, in part because it is used to stitch together a wide variety of political, economic, environmental, and social projects and experiences, and in part because it is increasingly used in a pejorative sense as dismissive shorthand by critics. In the most general sense, the term refers to an economic and political philosophy that questions, and in some versions entirely rejects, government interventions in the market and people's relationships to the economy, and eschews social and collective controls over the behavior and practices of firms, the movement of capital, and the regulation of socio-economic relationships. More dramatically, it is philosophy that describes itself in terms of "hard realism" but is often wrapped in a cloak of remarkably utopian promises, offering a world liberated to "unleash" the emancipatory power of markets and local decision-making.

This puts it in contradistinction not only with the concepts underpinning planned and socialist political economies but even with more Keynesian liberal politics and philosophies of the early and mid-twentieth century, which stressed controls, tinkering, and harnessing of markets within broader practices and goals of governance. As is becoming evident however, this (over-) simple definition of the phenomenon, while a useful starting place, detracts from many complex and contradictory elements, as recent discussions of neoliberalism's many forms and practices suggest.

In recent years, there has been a proliferation of critical literature on neoliberalism and neoliberalization, including an explosion in scholarship within geography. This includes early work on neoliberalism as post-Fordist regulation (Jessop 1994; Peck and Tickell 1995; Tickell and Peck 1995), neoliberal reforms at the urban scale (Brenner and Theodore 2002; Keil 2002; Swyngedouw *et al.* 2002), neoliberal reforms of labor markets (Peck 2002a, 2002b), the spatiality of neoliberal reforms (including scalar dimensions) (Peck and Tickell 2002), neoliberalism as development orthodoxy (Peet and Hartwick 1999; Peet and Watts 1993), and the nexus of neoliberalism and geopolitics (Harvey 2003; Sparke 2006). It also includes work at the relatively macro-scale of reforms to state–market relations as well as the micro-scale embodiment of the ways in which neoliberalism re-works citizenship and subjectivities (Mitchell 2003; Rankin 2001). And, most central to this volume, it includes the emergence of a distinct literature on the convergence of neoliberalism and nature, some of which is represented and re-worked here.

As a body, this scholarship explores various dimensions and arenas in which a range of diverse ideologies, discourses, and practices have been put in place as policies and institutions, and in the process have re-worked previous relationships between the market (e.g. commodities to be bought or sold and institutions facilitating trade), the state (e.g. bureaucracies managing exchange, labor, and the environment), and civil society (e.g. organizations assembled for collective good or defense of interests). This has included considerable empirical work on the way neoliberalism often circulates as lofty and undefined ideals, panaceas absent any reference to context ("thin policies"), but with effects entirely concrete and drastic in the lived experiences of human and non-human alike ("hard outcomes") (Peck 2001).

All this work also has inspired, perhaps inevitably, concern that the concept may be too big or bloated to capture the diversity of projects labeled as neoliberal. And in charting the historic and geographic changes that mark neoliberal transitions, concern has also been raised that distinctions such as "rolling back" and "rolling out" neoliberalism (Peck and Tickell 2002) are inadequate to convey the temporal and spatial evolution and adaptation of governance projects that, even if inspired by neoliberal orthodoxies, are transformed in the messiness of politics, lived experiences and actual geographies. While this diversity underpins the explicitly plural notion of "neoliberalizations," as well as efforts to theorize "hybrid" neoliberalisms (Larner 2003; McCarthy 2005, 2006), Barnett (2005) goes so far as to suggest that the word may now act as little more than a consolatory shibboleth for left-leaning academics, blinding them to the diverse experiences and pluralities of political struggles over governance.

There is a real danger here. No doubt, some invocations of "neoliberalism" act as little more than shallow, ahistorical and ageographical invocations of a poorly defined abstraction, perhaps ironically reinforcing the taken-for-granted character of free-market discourses and the typically idealist ways in which they are championed. We are acutely aware of this danger. How to capture important continuities and connections in diverse transformations going on around the world over a period of many years is no simple task, and any effort to do so requires observers to engage in some degree of abstraction, and to actively decide and disclose what in the particular cases they study is central rather than peripheral, necessary rather than contingent. Inevitably, the term becomes stretched as a result.

And yet something is going on worth naming. We insist on the analytical and political purchase of identifying, albeit reflexively, the pervasive "metalogics" of what we see around us, not least because we see a fairly common set of discourses, ideologies and practices that remain the most dominant development in social regulation in the post-Keynesian era. In our view, what is required is not to jettison the term, but rather to work the concept carefully through what Sparke (2006) calls "context-contingent analyses." The chapters included here comprise such analyses, focused on both the diverse and common qualities of recent reforms and the politics and ecology of environmental change. That said, in a general sense, what do we mean when we say "neoliberalism"?

At the most general or over-arching, "big-picture" level, neoliberalism has been examined by David Harvey as a global project to restore, renew, and expand the conditions for capital accumulation and, in related fashion, to restore power to economic elites (or to establish it where it did not already exist) (Harvey 2005; see also Duménil and Lévy 2004). Harvey argues that neoliberalism is not only an abstract set of ideas about how to best to organize society to facilitate the production of wealth and allow for the maximization of freedom, as many proponents of neoliberalism would have it. Rather, he argues that neoliberalism is an intensely political project, one in which economic elites more or less intentionally seek to increase their wealth and income, but also their political and economic freedom and flexibility by rolling back the redistributive reforms of the mid-twentieth century (particularly those adopted in the aftermath of the global Great Depression of the 1930s), reforms often dubbed by an additional shorthand concept of Keynesianism. As evidence, Harvey and others point to the fact that central elements of the neoliberal era have featured the rollback of regulations on capital accumulation coupled with reductions in social safety net provisions and state-coordinated redistribution of wealth and income, with evident consequences in spiralling social inequality.

While we agree in substantial measure, it is also important to recognize that neoliberalism has an important intellectual, discursive, and ideological lineage which helps lend at least the appearance of coherence and consistency to what we admit (and seek to problematize in this volume) as disparate, context-contingent projects. Returning to and fusing ideas drawn from classical political as well as economic liberalism (hence the name), neoliberal discourses tend to emphasize at least the idea (often selectively invoked in practice) of so-called "*laissez-faire*" economic regulation, i.e. shifting and "rolling back" the state apparatus where it is seen to impinge upon capital investment, commodity production, and market exchange, typically via championing abstract constructions of yeoman entrepreneurial capitalists and small businesses (as opposed to powerful, footloose multinationals) struggling under the oppressive weight of an overbearing state. Neoliberalism tends also to reinforce and celebrate strong private, individual, and exclusive property rights. Proponents tend to invoke specifically political notions of liberalism with emphasis on the rights, freedoms, and responsibilities of individuals, again typically posed in relation to a monolithic state represented as singularly in opposition to the realization of individual freedom.[2]

The intellectual lineage of neoliberalism is complex and in some ways contradictory, but certainly draws on a specific fusion of post-World War II economic and political liberalism forcefully articulated by the likes of Friedrich Hayek, Milton Friedman and Richard Epstein. In fact, arguably what marks neoliberal discourses and orthodoxies as a distinct body of social theory is a potent but polyvalent conflation of political and economic liberalism, one following from and arguably reducible to the promises of political emancipation through economic growth, increasing prosperity, and market mediated social relations. It is hardly coincidence that such a virulent and utopian articulation of ideas that link *laissez-faire* capitalism, liberal democracy, and individual freedoms would emerge and enjoy wide circulation—particularly in the US and the UK—during the Cold War period.

Still, many (including Harvey as noted above) link the actual institutionalization of neoliberalism to the period since the early 1970s, a period actually coinciding with declining profitability in leading capitalist economies (Harvey 1989; Arrighi 1994). It is primarily during this period that neoliberal practices were introduced as macro and micro-level reforms in response to a perceived crisis in so-called Keynesian social regulation, featuring an array of by now familiar devices:

- regressive reforms of state taxation and rollbacks in redistributive spending;
- privatization of services formerly provided by and through the state;
- reinforcement and extension of private, exclusive, and individuated property rights;
- liberalization of state regulations specifically governing trade and investment across international borders, though in uneven and contradictory ways that reflect not only the ideology of free trade, but also the political interests negotiating often highly selective and confusing blends of liberalization and protectionism;
- emphasis on state austerity and fiscal retrenchment with an associated defunding or outright cancellation of a wide array of social services, but again, in contradictory ways that are often combined with entrenchment of so-called supply side spending, e.g. development projects, programs to support economic innovation and competitiveness, and of course military investments;
- workfare, and other incentive-based schemes aimed at disciplining workers and civil servants (and at least nominally at increasing productivity and efficiency), accompanied by deregulation and re-regulation of labor markets;
- the restructuring of state regulatory apparatuses in ways that tend to enhance private and corporate authority over economic, environmental, and social action;
- offloading and decentralization involving both the re-scaling of governance up and down from nation-states, as well as the recruitment of volunteer, civil society-based organizations to undertake many functions formerly provided by states.

These and other strategies characterize the institutionalization of neoliberalism proliferated via a spate of interconnected reforms of governance that have swept through most of the world, particularly since the 1980s.

Within this lineage, some argue that important systemic shifts have occurred, including a transition from so-called "rollback" reforms to subsequent, Third-way style "roll-outs" (Peck and Tickell 2002). Others note distinctions between the pure market orthodoxy pursued by neoliberal advocates and the diverse, hybrid forms that these changes tend to comprise when filtered through the complex apparatus of social and political life (Jessop 2002). Such distinctions recognize diversity in the ways that neoliberal ideas take hold in and respond to different contexts, evident, for instance, in important differences between reforms in rich countries as opposed to variants formulated as development policy in the era of the debt crises and propagated in poor countries via the vehicle of structural adjustment plans (SAPs). While the diversity evident in these different kinds of changes does indeed pose challenges for agreeing on any singularity called neoliberalism, it is at the same time the almost taken-for-granted character of the list of strategies listed above – their truth-like obviousness and familiarity – that testifies to the deep hold of certain repeated tropes on the contemporary political imagination as the only common-sense or natural (and naturalized) ways that realistic people can think or imagine policy prescriptions. Neoliberalism is in this sense a predominant set of beliefs about the world.

Yet it bears stressing that neoliberalism has come to occupy this dominant ideological position not primarily through any "inherent" power of the ideas themselves, but rather through political mechanisms and institutions that propel them to travel and become entrenched. Organizations such as the International Monetary Fund and the World Bank (along with their financial backers) have for some time insisted that countries borrowing money from them adopt strongly neoliberal policies, not least using SAPs as instruments. Ideas also travel through networks and gatherings of powerful professionals and agenda setters, e.g. the World Economic Forum (an annual gathering of economic and political elites in Davos, Switzerland). Successive rounds of elite level, generally non-accountable and non-transparent negotiations over trade liberalization, via the GATT and the WTO, as well as in regional trading blocks such as the North American Free Trade Agreement (NAFTA) have also established important precedents for the neoliberalization of trade and investment, including by dramatically enhancing the rights and powers of private investors (Clarkson 2003; McCarthy 2004).

And while this traveling and stretching of policies and projects helps consolidate neoliberalism as the dominant post-Keynesian mode of regulation, it also leads to important particularities whose geographies matter. This is in part due to the ways in which neoliberal reforms have arisen out of and in response to highly particular crises in governance (Jessop 2002). In Chile, the authoritarian brutality of the Pinochet coup and subsequent regime change swept away an incipient state socialism, providing the impetus and the muscle behind sweeping changes championed by the so-called "Chicago Boys" (high-level Chilean economic advisors), educated in neoliberal orthodoxy at the University of Chicago (Schurman 1996; Valdes 1995). In mid-1970s New York City, the city's financial ruin provided opportunities for crisis-propelled, highly revanchist reforms to be introduced (not least by a class of leveraged financial elites) in the name of rescuing the city from financial ruin (Harvey 1985, 2005; Smith 1996). That the particularities of rules and laws should vary in these two cases, along with the institutions, stories, motivations, responses, and effects of these reforms, should in no way be surprising, given the deeply contextual conditions of their invention and implementation.

In the wake of the elections of Ronald Reagan in the United States and Margaret Thatcher in the UK, neoliberal reforms were propelled in the context of persistent economic recession, inflationary pressures that undermined both consumption and investment, and a perceived threat that the national economies of the USA and the UK were losing their competitive advantages vis-à-vis Japan and other newly industrializing nations. In these instances, national economic competitiveness was mobilized as a discourse tethered to the fate of individual citizens in ways that enlisted broad support, and disciplined citizens for the sacrifices necessary to restore profitability and growth (Hall 1988). The immediacy of this crisis-driven reform was infamously summed up by Margaret Thatcher when she declared in the UK context that "There is no alternative," which has since become pejoratively invoked as the coercive catch-phrase reflecting the commonsensical and hegemonic power of neoliberal discourses.

Crisis-inspired reform is also evident in the threatened bankruptcy of Mexico's national economy in 1982, the first in a series of so-called debt crises that swept through largely post-colonial economies of the global south during the 1980s and 1990s. Acute financial pressure on heavily indebted nations created the leverage for multilateral lenders and private banks to propagate neoliberal development orthodoxy as the so-called "Washington consensus," accelerated by dissolution of the Soviet Union and the end of the Cold War (Escobar 1995). These disparate cases, tethered together by common threads, speak to the ways in which the highly idealist, utopian promises of neoliberal discourse become territorialized in specific places at specific times.

Yet, if disparity in experience is evident in even the earliest neoliberal experiments, projects, and ideas, and if this diversity has only become more evident by a growing literature on geographically diverse contexts, does this warrant disposal of the term? Has "neoliberalism" become merely a consolatory bookmark for leftist critics (comparable to "secular humanism" on the right) rather than a name for something tangibly evident and worth worrying about? We obviously argue no, as do activists around the world who recognize and seek to name something they viscerally recognize. A singular neoliberalism can only be sustained at the cost of acknowledging that it "averages out" important differences between disparate political economic and ecological projects and experiences. But this is in part a consequence of the fact that what is invoked as "neoliberalism" is far more than mere sloganeering and that the stakes in naming and recognizing it are high.

Does this mean scholars should blindly follow the political slogans of social movements invoking neoliberalism as their *bête noire*? Hardly, but it does mean we should take these invocations seriously, and in the interests of aligning ourselves with but also advancing (not merely following) these movements, we ought to reflexively consider the particular valences of neoliberalism in the context of specific projects, evaluate how the cases we examine illuminate and help us to understand both difference and continuity, and reflect on the changing character of reforms, themselves arising out of political struggle (not merely in abstraction). When the time comes to jettison the term because it can no longer meaningfully characterize developments in the world around us, and just as importantly, because the term no longer does adequate political and intellectual work (a time that will surely come), this will happen through the complex interplay of politics and scholarship, not through wordplay. In the meantime, we are left with the challenge, as Castree (2005; 2006 and in this volume) rightly observes, to make sure we are reading one another's work carefully in looking for ways to make commensurable the lessons we learn from diverse studies. We must embrace the challenges to this commensurability that arise from the way in which the politics, the discourses, and the material circumstances of our respective neoliberalizations are related but different. But the burden falls on us as scholars (and activists) to seek out these connections and common problems, make them plain and compelling to a broader public, and subvert the sense of inevitability that surrounds them.

These are central goals of this volume. The contribution of cases is therefore meant to portray diverse struggles and misadventures of neoliberalizations, but also to point the way towards common experience. The collection reflects our commitment, shared by the contributors, that engagement with concrete political ecological circumstances is vital to critique, and that far from militant particularisms, these cases are the starting point for moving past neoliberalism's taken-for-granted quality. We want the cases to be read against one another, and we sincerely hope that readers, abetted by the commentators, will find ways to make them commensurable.

The nature of neoliberalism and the neoliberalization of nature

This brings us to the second of our introductory questions, concerning the specifically environmental dimensions of neoliberalism. This volume is born out of our diverse experiences, shared with many of the contributors, in trying to explore not only the environmental impacts of neoliberal reforms (as important as that remains), but also to consider the ways in which environmental governance, and environmentalism as a set of political movements, coincide, collide, articulate, and even constitute the emergence of neoliberalism.

In this sense, we view the relationship between neoliberal reform, on one hand, and environmental politics, governance, and change, on the other, as more than coincident. Rather we suggest that they are inherent in the consistent imperative that runs through the history of neoliberalization: to expand opportunities for capital investment and accumulation by re-working state–market–civil society relations to allow for the stretching and deepening of commodity production, circulation and exchange. When this is combined with a stress on individual rights and freedoms, especially private property rights, there is a necessary re-working of the way human society and non-human systems and beings relate. Indeed, this is hardly new or surprising, in as much as traditional liberalism in its various guises was in part a product of, and propelling force for, historic efforts in the re-working of socio-nature (McCarthy and Prudham 2004).

Consider, for instance, the way biophysical nature in the present day (including the human body) is an important frontier for the expansion and deepening of commodification. As Kloppenburg (2004) and Wright (1994) each show in their own way, this dynamic is central to the history of bio-technology, initially centered in the US and the UK, and increasingly expanding to the global scale. Genes, genetically modified organisms and other products of the new biotechnology have been explicitly targeted since at least the early 1970s in both the US and the UK as important arenas for the expansion of capital accumulation in agriculture and health sciences, not to mention other spheres.

Yet this "new" frontier is arguably only a continuation of a more deeply historical process. John Locke's moral defense of individual property and his early articulation of *laissez-faire* logics were both fabricated in, and justified by, the crucible of enclosure, that process in which complex assemblages of communal rights to nature were terminated in England and elsewhere in favor of private ownership (Locke [1690] 1952) [#2289]. In this way, neoliberalism and liberalism are both products of, and drivers toward, reconfigurations of socio-natural systems. In both cases, moreover, transition is neither inevitable or smooth, requiring as it does coercions, political contests, physical confrontations, and deliberate manipulations of institutions, including in science policy, state–industry–university relations, and property rights.

All of this has culminated in expanded and highly contested opportunities for capitalist profit-making in the production of life, moreover. It has also become an increasingly global project, as biotech firms (typically championed by the US government) press for adoption of strong, individuated, exclusive private property rights over life forms (McAfee 2003). Our point here is that this history is hardly a mere manifestation of pre-conceived neoliberal ideas, but rather it is among the earliest examples of the instantiation of neoliberalism born out of a crisis of profitability, solved ostensibly by recourse to changes in our relationship to the more-than-human world. Parallels may be found in the ways in which the restructuring of socio-natural relations are often central to the specifics of Structural Adjustment Plans as they target the liberalization of investment and trade restrictions in nature-centered sectors such as agriculture and forestry, rolling back the state regulatory apparatus, and mandating the privatization of water utilities and service provisioning (Clapp 2000; Kincaid 2001). As these sectors and ecologies are transformed by previous generations of economic activity, they provide opportunities for new markets and systems of extraction, which in turn lead to new environmental outcomes. Neoliberal reform is both a cause of environmental change and a product of changes in the way we interact with the environment.

At the same time, neoliberalization and environmentalism (as a movement and a practice of governance) also share a complex and conjoined lineage in the past few decades. As two of us have argued elsewhere (McCarthy and Prudham 2004), environmentalism paradoxically points to the politically contested character of neoliberalism but also to the pervasive, taken-for-granted and apolitical appearance of neoliberal discourses and practices. In the early years of the Reagan era, for example, important initiatives aimed at scaling back environmental regulatory safeguards and standards became focal to the institutionalization of American neoliberalism (Dryzek 1996; Vig and Kraft 1984). Yet these also became focal points for resistance, as environmental social movements across the country organized against regulatory rollbacks. This speaks to the salience and power of environmentalism in America, and elsewhere, and to one of the more apparent roadblocks to the grand ambitions of neoliberal reformers. Yet, the assault on standards, funding for environmental programs and remediation, and attempts to remove restrictions on capital's access to nature has remained a focal point in American neoliberalism right up to the present day, taking several forms. This includes non-action such as the current Bush administration's stonewalling on climate change. It also includes more overt rollbacks of the sort that helped structure Hurricane Katrina's disastrous implications for New Orleans, retreats whose organized irresponsibility was revealed in terms of highly racialized urban social polarities and environmental injustice only in the aftermath of the disaster.[3]

But if environmental change and environmental politics have become arenas for debating the limits, costs, and consequences of neoliberalizations, they have provided equal evidence of the power of neoliberal orthodoxies to circulate through and hybridize with environmentalism, comprising part of what makes neoliberalizations complex and variegated. This is evident, for example, in the internal debate among powerful environmental groups over the role of market-based incentives and mechanisms in environmental governance. Mechanisms institutionalized via the Kyoto Protocol, pushed and negotiated in substantial measure by NGOs, helped construct and produce a global community of participation in tradeable, commodity-like carbon permits and offsets. This follows important alliances forged in the United States around Clean Air Act tradeable permit schemes aimed at reducing a variety of pollutants at the urban scale, and points to the capacity of environmental organizations not only to resist, but also to adapt to and help spread neoliberal-inspired approaches to governance. So too it includes rollouts into environmental policy, where market mechanisms are introduced into environmental regulation, e.g. in contemporary schemes targeting wetland banking in the United States (see Robertson 2004 and this volume). That these measures tend to entrench a utilitarian and fetishistic disposition toward the biophysical world does not seem to trouble those groups willing to pursue clean-up and preservation at any cost.

In similar fashion, groups such as the Nature Conservancy have sought to incorporate the privatization of nature as a tool, and enlist its proponents as allies, by recruiting income to buy up tracts of land thought valuable for the preservation of biodiversity and rare habitats. If these measures are not exactly aimed at expanding opportunities for capital accumulation, they still pose interesting and largely unexamined ways of locking up surplus capital (often highly sheltered from taxation) in the form of socionatural fixes, while doing nothing to subvert the hegemony of highly individualistic and exclusive property rights over nature. Internationally, parallels may be drawn in the highly utilitarian, option value constructions that surround efforts to enclose and protect biodiversity, in ways that can be highly elitist, exclusionary, and neo-colonial. Efforts to enclose, and render exchangeable, hunting rights for the Ibex in northern Pakistan in the name of conservation represent a particularly blatant and galling example (MacDonald 2005).

Linking the question of nature to neoliberalizations is, for these reasons, more than just a matter of analyzing economic policy impacts on the environment. Rather, it both promises to disclose important ways that specific, potentially pernicious ideologies and policies are in part compelled and constituted through our changing relationship to nature, and to point to environmental politics and environmental governance as key arenas for extending and hybridizing political and economic projects.

We also argue that engaging with environmental neoliberalism is a critical strategy for understanding these changes and outcomes as context-specific and constituted both by discourses and material conditions and processes. This reflects learning over the past few years in the area of political ecology, a signature field of interdisciplinary scholarship. If the question of environmental change and its politics is central in this field, so too is the idea that understanding environmental change requires analysis across scales, from the highly local ways people make and represent ecological processes to the ways in which their experiences and understandings are nested in complex networks of interacting biophysical, institutional and discursive processes (Robbins and Fraser 2003; Turner 1993). Many of the contributors to this volume have been informed by the emergence of political ecology, and this is no accident. The theoretical and methodological commitments to grounded engagement with actual places, people, and ecologies in political ecology provide a powerful way to check the idealist tendencies of neoliberal discourses and ideologies.

So too, political ecology forces an appreciation of the way such discourses and ideals must be negotiated in relation both to socially constructed knowledge of the biophysical world (e.g. local agro-ecological knowledge) as well as the irreducibly material character of that world (e.g. local rates of soil erosion). Because of this, neoliberalizations of resource and environmental management must be negotiated and concretized in relation to biophysical processes, as well as human uses and understandings of these processes, making for complex outcomes. The attempted development of individual, transferable quotas in fisheries, where the resource in question is fugitive, unpredictable and has complex biotic and abiotic feedbacks, is a prime example (see Mansfield 2004 and this volume).

Finally, political ecology opens the door on understanding how technical discourses of environmental conditions or change (e.g. deforestation or desertification) are enrolled in political and economic momentum for enclosure, control, and reconfiguration of socio-nature. This raises difficult questions concerning the relationship between science, power, economy, and society. Such questions are better faced head-on, however, rather than avoided by relegating environmental science to a realm somehow separate or isolated from economics or politics (Robbins 2004; Heynen *et al.* 2006).

The world is not an isotropic plane, as geographers ought to know well, and the uneven geography of socio-nature is one of the reasons why not. But we maintain that appreciating the inherently context-contingent and material-discursive character of neoliberalizing nature is not just about doing good political ecology. Rather, this is an important arena in which the utopian idealism of neoliberal discourses and ideologies may be evaluated against concrete experiences and outcomes. In this sense, experiences with neoliberal environments in our view go a long way toward challenging and disclosing the (false) promises of neoliberalism more generally.

Organization of this volume

All this goes some way toward arguing for the reality of something called neoliberalism, its relationship to nature, and the character of critical study required to dissect it. Even so, the diverse components of the larger whole can be understood on their own terms, As a result, the collection is organized into four parts, each of which addresses one such thematic component, includes multiple case studies, and provides unifying commentaries. These sections are, in order from first to last, "Enclosure and Privatization," "Commodification and Marketization," "Devolution and Neoliberal Governmentalities," and "Resistance." This grouping of the essays is intended to accentuate what we see as four key underlying themes in the collection. It is not meant to suggest an exclusive division of substantive and conceptual labor, and indeed, much commonality exists across chapters grouped in different sections. Similarly, while the commentaries specifically respond to the chapter groupings, there are important cross-cutting themes.

Part I, "Enclosure and privatization," is a collection of chapters dealing broadly with the re-working of property relations governing access to and control of nature under neoliberal reforms. These reforms are generally in the direction of greater individuation, exclusivity, private control, and (in many instances) marketization, typically legitimated and represented in terms of enhancing individual freedoms, economic efficiency, and environmental quality. Paul Robbins and April Luginbuhl in Chapter 1 examine the quixotic and politically and ecologically uneven development of private hunting rights over wildlife in the American West. In Chapter 2, James McCarthy interrogates the ways in which environmental quality and ecological functions (which he theorizes as conditions of capitalist production) become focal in dramatic shifts in political and property rights as administrative jurisdiction shifts from national to supranational, and from public to private, in the ongoing institutionalization of the North American Free Trade Agreement. Erik Swyngedouw in Chapter 3 examines the question of accumulation by dispossession head-on and in contrast to "normal" capitalism, arguing on the basis of his wide ranging summary of the privatization of water rights that these schemes must indeed be seen as polyvalent dispossessions. Becky Mansfield in Chapter 4 situates the development of neoliberal-inspired property reforms over fishing rights against a backdrop of a longstanding struggle over a socially negotiated construction of the commons question in fisheries, i.e. the persistence and persistent problematization of open access (conflated with communally held) fishing rights in the face of various schemes to "enclose the ocean." The last chapter in this section by Gavin Bridge concerns the creation of private gold mining rights in Guyana. Bridge converges with Mansfield in locating recent privatization schemes against a longer historical political economy and ecology of enclosing mining rights, even as neoliberal schemes literally deepen the hold of exclusively private rights of access in the mining sector. In their commentaries, both Nancy Lee Peluso and Jim Glassman interpret the chapters in this part in light of current debates regarding contemporary primitive accumulation, highlighting the empirical variety of its environmental manifestations and their potential to create barriers to neoliberal projects. Peluso focuses on the enduring importance of states and nations to neoliberalism, emphasizing that all the cases on offer here require state re-regulation rather than de-regulation; her central question is why states and their citizens accede to such such transformations and reterritorializations. Jim Glassman offers, indirectly, at least one possible answer to Peluso's question, advocating the view that neoliberalism ought to be understood largely as a class project of the most powerful and mobile capitalists, who have dramatically increased their ability to move their assets relatively freely among states and territories.

Part II, "Commodification and marketization," brings chapters together that collectively illustrate the political economic processes through which the environment broadly, and particular ecosystems more specifically, are reduced to commodities through pricing mechanisms that open them up to free-market profiteering and often time destruction. The chapters in this part show how revolutions in law, policy, and markets are accelerating the ongoing commodification of natural things, laying bare the structurally-driven and environmentally-destructive tendencies of capitalism. First, in Chapter 8, Karen Bakker asks questions about market environmentalism in the water supply in England and Wales, including what have been the impacts of re-regulation of water; what is the analytical utility of the term "neoliberalism" in describing these changes; and can the project of water supply privatization and re-regulation be categorized as a success, and on what grounds? Next, in Chapter 9, Morgan M. Robertson discusses wetland mitigation banking and problems in environmental governance related to how banking complicates "smooth" neoliberal discourse about the process of commodifying ecosystem services. Gail Hollander in Chapter 10 discusses agricultural trade liberalization both via the language of EU agricultural trade advocates and as it has played out in the south Florida landscape for the sake of illustrating how ideas of landscape, livelihood and agro-ecology, encompassed by the term "multifunctionality," have been used in defense of domestic agricultural supports. Douglas Young and Roger Keil close Part II with a chapter that shows how growth in the Greater Toronto Area has had a significant impact on, and articulation with, water and how the process of privatization at the core of this growth has exploded as a result of a less regulated and market-driven global economy. The commentary by Neil Brenner and Nik Theodore, researchers who have made major contributions to theoretical understandings of neoliberalizations in other domains, situates the chapters and the volume as a whole on a wider canvas of attempts to theorize neoliberalism writ large, clearly delineating many of the key issues. Reinforcing one of the guiding principles of this volume, they reiterate the necessity of concrete research on what they term "actually existing neoliberalisms."

In Part III, the substantive focus shifts to "Devolution and neoliberal governmentalities," where various experiments in shifting responsibility, accountability, and management abandon or reconfigure state controls over nature. These cases show the extremely complex outcomes that can ensue in devolution, showing the "state" from very different points of view. As Heynen and Perkins in Chapter 15 set their viewpoint from the position of minority communities to show the structured nature of local state planting and neglect of urban forests in Milwaukee, Holifield's account of Superfund in Chapter 16 is narrated from within the EPA itself, showing the uneven relation of governmentality from within a real entity we typically understand as "government." So too, as only case study research can show, things often turn out differently in the process of devolution than one might expect. While Prudham's grim exploration of water quality crisis in Walkerton, Ontario, in Chapter 13 shows the very real and material negative outcomes of governmental experiments, in Chapter 14 Blomley's urban gardens in Strathcona, Vancouver, show how planned exclusions become inclusive possibilities. Together they paint a picture of local actors negotiating neoliberal abstractions and conceits about the state in very concrete ways. The commentaries by Wendy Larner and Dianne Rocheleau both focus on the complexities and possibilities hidden by and perhaps even latent within relatively monolithic conceptions of neoliberalism: Larner suggests that the authors underestimate the compatibility of neoliberal reforms with a wide range of political configurations and outcomes, while Rocheleau emphasizes that neoliberalism's production of new subjectivities and institutions is itself a likely source of resistance and change over time.

The fourth and final part explores the prospects for "Resistance" in this arena, examining both some of the many ways in which neoliberal environmental measures have been stalled, reversed, or turned to unexpectedly progressive ends by activism of multiple sorts, and how to theorize the always complex imbrications of resistance and domination under any sort of liberal regime. Examples include efforts to free public spaces and goods from enclosure, calls for decommodification, and the explicit politicization of neoliberalism and the championing of viable alternatives (both in direct opposition to attempts to present neoliberal policies as natural and necessary). Resistance is too often interpreted as a nearly deterministic result of a Polanyian double movement, but the cases show that resistance is as complicated and constrained by contingencies as neoliberalism itself. Both resistance to neoliberalism and its complexities are at the forefront of Wendy Wolford's study of land reform programs in Brazil during the neoliberal era in Chapter 20; the programs examined explicitly pit market-versus-society-centered conceptions of rights against each other, yet both are administered by the state and both appeal to complex moral economies for legitimacy. In Chapter 21, Kevin St. Martin focuses on the importance of alternatives, demonstrating how New England fishing communities continue to operate as commons in ways that both contradict and underpin the dominant theories and administration of the fishery. David Correia's examination of historical struggles over forests in New Mexico during a period prior to neoliberalism, in Chapter 19, meanwhile, illustrates that the struggles over the capitalization of nature were hardly absent from the Keynesian era, either, and that some of those struggles paved the way for what would later be presented as neoliberal "common sense." In her commentary, Juanita Sundberg emphasizes the importance of understanding alternatives and resistance through grounded, particular research, and of researchers studying and working with groups resisting neoliberalism, rather than maintaining a more conventional academic distance. Michael Watts, finally, emphasizes the need to understand the complex genealogies of neoliberalism itself in order to mount or analyze effective resistance to it.

Taken together, there are barriers to finding commonalties in the cases, to be sure. The first barrier is evident in the variety of settings and outcomes for neoliberal reform, which merely confirm the heterogeneity of various neoliberalizations. There is no singular neoliberalism evident here. Complex, contingent, and contextual political and cultural influences, together with diverse material ecologies and economies comprise very different settings for reforms in ways that shape and constrain their direction. Nor do the authors conceptualize neoliberalism in a uniform manner, a problem endemic to a collection such as this, all the more so since the research was not undertaken collectively. Ultimately, then, not all the authors arrive at the same normative conclusions. Robertson, for instance, finds little to celebrate in the development of exchangeable wetlands, while Bakker is open to the notion that privatization of water might not be inevitably anathema to improved quality.

At the same time, however, there are common threads in both approaches and conclusions. First, on balance, the cases cast serious doubt on utopian predictions such as those of CATO's Jerry Taylor, promising a bright world of green capitalism to be achieved via neoliberal reform. The evidence in these chapters shows reason for concern, and the reviews are at best mixed, pointing to the need to consider carefully the specific character of both governance reforms and social and environmental outcomes in geographical context. In worst case scenarios, the results of ill-conceived market-based reforms have proven disastrous either socially, ecologically, or both. Where neoliberal reform has provided social and ecological opportunity, it has often done so in spite of, and not as a result of, enclosure, commodification, and devolution.

Significantly, however, while most if not all of the contributors here are skeptical about the general thrust of environmental neoliberalism, and though their specific conclusions are diverse, they also share a common cognizance of the need to move beyond polemics and to identify political openings. This means carefully considering the opportunities and limits of non-governmental organizations as many step into the breach left by downsized or hollowed-out states, including the ways in which resource users assert and constitute themselves communally in the midst and in spite of efforts to individuate and privatize (St. Martin 2005, and this volume). So too, it means examining the mix of regulatory measures and oversight that accompanies (or fails to accompany) privatization and administrative offloading (Prudham 2004; Rees 1998; Smith 2004).

More generally, it means using common case experiences to point the way to alternatives and opportunities. By showing the emergence and implementation of neoliberal reform as an ideology, with a history and a purpose, we assert that it is possible to subvert its taken-for-granted political status. Many of us (with some dismay!) have witnessed the emergence of this status in our teaching, as students interpret Hardin's parable of already individuated and selfish herders contending for access to a pasture with fixed carrying capacity as an accurate, ahistorical and ageographical rendition of political ecological reality, a simple truth. We are keen to disrupt this truth-like status, but to avoid making it worse with totalizing or monolithic narratives in the ways in which we represent neoliberalism. The theoretically informed but empirically oriented character of these cases is no accident, but is instead meant to help authors and readers alike work through specifics, to pose questions about pre-given notions, and to consider alternatives both foreclosed and yet to be explored. The stakes are high and rigorous research and theory remain central tools for imagining, asserting, and implementing progressive and sustainable futures. We hope this collection contributes to this effort by bringing diverse experiences together.

Notes

1 We use the term "neoliberalization" frequently, drawing most directly on discussions by Peck (2001) and Peck and Tickell (2002), to capture something of the diversity of different neoliberal-inspired projects in governance reform, not least in the diversity of ways that neoliberal tropes are territorialized. As Peck and Tickell (2002) write: "we propose a processual conception of neoliberalization as both an 'out there' and an 'in here' phenomenon whose effects are necessarily variegated and uneven, but the incidence and diffusion of which may provide clues to a pervasive 'metalogic'." Whether this gets around the tricky ontological but also important political question as to whether the singularity of "neoliberalism" obscures more than it discloses is a subject we discuss at more length below.

2 That states can also be a source of emancipation, including at individual levels is typically unexplored terrain in neoliberal discourses, even though state sanction of individual property rights is one manner in which individual freedoms are pursued *through* not in opposition to states. Similar logic might be applied to state governed liberal notions of citizenship that act in important ways to shape and constrain individual opportunities.

3 Peck (2006) also notes that the debate over Hurricane Katrina is indicative of the ways in which highly idealized, utopian representations of market orthodoxy are invoked in relation to real world political ecologies like this one in order to argue – somewhat perversely – that failures originate in *not going far enough* with *laissez-faire* reforms!

References

Arrighi, G. (1994) *The Long Twentieth Century: Money, Power, and the Origins of Our Times*, London: Verso.

Barnett, C. (2005) "The Consolations of Neoliberalism," *Geoforum* 36: 7–12.

Brenner, N. and Theodore, N. (2002) "Cities and the Geographies of 'Actually Existing Neoliberalisms,'" *Antipode* 34(3): 349–79.

Britto Garcia, L. (2006) "Interview with Venezuelan Author Luis Britto Garcia: The Fourth World War is Against Neoliberalism," *Terra Viva*, January 28, available at: www.ipsterraviva.net/tv/wsf2006/viewstory.asp?idnews=541

Castree, N. (2005) "The Epistemology of Particulars: Human Geography, Case Studies, and Particulars," *Geoforum* 36: 541–44.

—— (2006) "Commentary: From Neoliberalism to Neoliberalisation: Consolations, Confusions, and Necessary Illusions," *Environment and Planning A* 38(1): 1–6.

Clapp, J. (2000) "The Global Economy and Environmental Change in Africa," in R. Stubbs and G. Underhill (eds) *Political Economy and the Changing Global Order*, Oxford: Oxford University Press.

Clarkson, S. (2003) "Locked In: Canada's External Constitution under Global Trade Governance," *American Review of Canadian Studies* 33(2): 145–74.

Dryzek, J. S. (1996) *Democracy in Capitalist Times: Ideals, Limits, and Struggles*, New York: Oxford University Press.

Duménil, G. and Lévy, D. (2004) *Capital Resurgent: Roots of the Neoliberal Revolution*. Cambridge, MA: Harvard University Press

Escobar, A. (1995) *Encountering Development: The Making and Unmaking of the Third World*, Princeton, NJ: Princeton University Press.

Hall, S. (1988) "The Toad in the Garden: Thatcherism among the Theorists," in C. Nelson and L. Grossberg (eds) *Marxism and the Interpretation of Culture*, Chicago, IL: University of Illinois Press, pp. 35–57.

Haraway, D. J. (1991) *Simians, Cyborgs and Women: The Reinvention of Nature*, New York: Routledge.

—— (1997) *Modest*Witness@Second*Millennium.Femaleman*Meets*Oncomouse: Feminism and Technoscience*, New York: Routledge.

Harvey, D. (1985) *The Urbanization of Capital: Studies in the History and Theory of Capitalist Urbanization*, Baltimore, MD: Johns Hopkins University Press.

—— (1989) *The Condition of Postmodernity: An Enquiry into the Origins of Cultural Change*, Oxford: Blackwell.

—— (2003) *The New Imperialism*, Oxford: Oxford University Press.

—— (2005) *A Brief History of Neoliberalism*, Oxford: Oxford University Press.

Heynen, N, Kaika, M. and Sywngedouw, E. (2006) *In the Nature of Cities: Urban Political Ecology and the Politics of Urban Metabolism*, London: Routledge.

Jessop, B. (1994) "Post-Fordism and the State," in A. Amin (ed.) *Post-Fordism: A Reader*, Oxford: Blackwell: pp. 251–79.

—— (2002) "Liberalism, Neoliberalism, and Urban Governance: A State-Theoretical Perspective," *Antipode* 34(3): 452–72.

Keil, R. (2002) "'Common-Sense' Neoliberalism: Progressive Conservative Urbanism in Toronto, Canada," *Antipode* 34(3): 578–601.

Kincaid, J. (2001) "Life and Debt." Film produced by Stephanie Black; distributed by New Yorker Video.

Kloppenburg, J. R. (2004) *First the Seed: The Political Economy of Plant Biotechnology*, 2nd edn, Madison, WI: University of Wisconsin Press.

Larner, W. (2003) "Neoliberalism?" *Environment and Planning D: Society and Space* 21: 509–12.

Locke, J. ([1690] 1952) *The Second Treatise of Government*, 1st edn, New York: Macmillan.

Lysandrou, P. (2005) "Globalisation as Commodification," *Cambridge Journal of Economics* 29(5): 769–97.

MacDonald, K. I. (2005) "Global Hunting Grounds: Power, Scale, and Ecology in the Negotiation of Conservation," *Cultural Geographies* 12: 259–91.

Mansfield, B. (2004) "Rules of Privatization: Contradictions in Neoliberal Regulation of North Pacific Fisheries," *Annals of the Association of American Geographers* 94(3): 565–84.

McAfee, K. (2003) "Neoliberalism on the Molecular Scale. Economic and Genetic Reductionism in Biotechnology Battles," *Geoforum* 34(2): 203–19.

McCarthy, J. (2004) "Privatizing Conditions of Production: Trade Agreements as Neoliberal Environmental Governance," *Geoforum* 35(3): 327–41.

—— (2005) "Devolution in the Woods: Community Forestry as Hybrid Neoliberalism," *Environment and Planning A* 37(6): 995–1014.

—— (2006) "Neoliberalism and the Politics of Alternatives: Community Forestry in British Columbia and the United States," *Annals of the Association of American Geographers* 96(1): 84–104.

McCarthy, J. and Prudham, S. (2004) "Neoliberal Nature and the Nature of Neoliberalism," *Geoforum* 35(3): 275–83.

Mitchell, K. (2003) "Educating the National Citizen in Neoliberal Times: From the Multicultural Self to the Strategic Cosmopolitan," *Transactions of the Institute of British Geographers* 28: 387–403.

Peck, J. (2001) "Neoliberalizing States: Thin Policies/Hard Outcomes," *Progress in Human Geography* 25(3): 445–55.

—— (2002a) "Labor, Zapped/Growth, Restored? Three Moments of Neoliberal Restructuring in the American Labor Market," *Journal of Economic Geography* 2(2): 179–220.

—— (2002b) "Political Economics of Scale: Fast Policy, Interscalar Relations, and Neoliberal Workfare," *Economic Geography* 78(3): 331–60.

—— (2006) "Liberating the City: Between New York and New Orleans." forthcoming in *Urban Geography.*

Peck, J. and Tickell. A. (1995) "The Social Regulation of Uneven Development: 'Regulatory Deficit', England's South East, and the Collapse of Thatcherism," *Environment and Planning A* 27: 15–40.

—— (2002) "Neoliberalizing Space," *Antipode* 34(3): 380–404.

Peet, R. and Hartwick, E. R. (1999) *Theories of Development*, New York: Guilford Press.

Peet, R. and Watts, M. (1993) "Introduction: Development Theory and Environment in an Age of Market Triumphalism," *Economic Geography* 69(3): 227–53.

Prudham, S. (2004) "Poisoning the Well: Neoliberalism and the Contamination of Municipal Water in Walkerton, Ontario," *Geoforum* 35(3): 343–59.

Rankin, K. (2001) "Governing Development: Neoliberalism, Microcredit, and Rational Economic Woman," *Economy and Society* 30(1): 18–37.

Rees, J. (1998) "Regulation and Private Participation in the Water and Sanitation Sector," *Natural Resources Forum* 22(2): 95–105.

Robbins, P. (2004) *Political Ecology: A Critical Introduction*, Malden, MA: Blackwell.

Robbins, P. and Fraser, A. (2003) "A Forest of Contradictions: Producing the Landscapes of the Scottish Highlands," *Antipode* 35(1): 95–118.

Robertson, M. (2004) "The Neoliberalization of Ecosystem Services: Wetland Mitigation Banking and Problems in Environmental Governance," *Geoforum* 35: 361–73.

Schurman, R. A. (1996) "Snails, Southern Hake, and Sustainability: Neoliberalism and Natural Resource Exports in Chile," *World Development* 24(11): 1695–1709.

Smith, L. (2004) "The Murky Waters of the Second Wave of Neoliberalism: Corporatization as a Service Delivery Model in Cape Town," *Geoforum* 35(3): 375–93.

Smith, N. (1996) *The New Urban Frontier: Gentrification and the Revanchist City*, London: Routledge.

Sparke, M. (2006) "Political Geography: Political Geographies of Globalization (2) – Governance," *Progress in Human Geography* 30(3): 357–72.

St. Martin, K. (2005) "Disrupting Enclosure in New England Fisheries," *Capitalism, Nature, Socialism* 16(1): 63–80.

Swyngedouw, E., Moulaert, F. and Rodriguez, A. (2002) "Neoliberal Urbanization in Europe: Large-Scale Urban Development Projects and the New Urban Policy," *Antipode* 34(3): 542–77.

Taylor, J. (2003) "Happy Earth Day? Thank Capitalism," *New York Sun*, April 22: Editorial and Opinion, p. 6.

Tickell, A. and Peck, J. (1995) "Social Regulation after Fordism: Regulation Theory, Neo-Liberalism, and the Global-Local Nexus," *Economy and Society* 24(3): 357–86.

Turner, M. D. (1993) "Overstocking the Range: A Critical Analysis of the Environmental Science of Sahelian Pastoralism," *Economic Geography* 69(4): 402–21.

Valdes, J. G. (1995) *Pinochet's Economists: The Chicago School of Economics in Chile,* Cambridge: Cambridge University Press.

Vig, N. J. and Kraft, M. E. (eds) (1984) *Environmental Policy in the 1980s: Reagan's New Agenda*, Washington, DC: Congressional Quarterly Press.

Wright, S. (1994) *Molecular Politics: Developing American and British Regulatory Policy for Genetic Engineering, 1972–82*, Chicago, IL: University of Chicago Press.

Part I
Enclosure and privatization

1 The last enclosure

Resisting privatization of wildlife in the western United States

Paul Robbins and April Luginbuhl

> Deer forests and the people cannot coexist ... one or the other must yield ...
> [Proprietors] follow a trade in deer with an eye solely towards profit ... The
> huntsman who wants a deer-forest limits his offers by no other calculation than
> the extent of his purse.
>
> (Robert Somers, "Letters from the Highlands: or the Famine of 1847"
> (as quoted in Marx 1967: 731))

> For little more than 20 years now an un-American movement has been trying to
> explode one of our nation's most sacred precepts, that of public ownership of
> wildlife ... The new Tories – mostly ranch owners with access to elk herds in
> these parts – increasingly are reserving their estates for a moneyed elite.
>
> (Ed Dentry, *Rocky Mountain News*, Sports Section, 1999)

Some of the earliest enclosures of public property were those that turned
wildlife into a commodity. As Somers described, via Marx, as early as 1847
deer were captured as an organism of profit, a form of "demurely domesticated cattle." Likewise, hunting, a prehistoric subsistence practice of the
Scottish Highlands, had been turned into an elite sport of accumulation.
Displacing smallholders, enclosed deer forests (private hunting estates containing "not a single tree") sprouted "like mushrooms" and "supplanted
sheep," driving farmers to "coarser" upslope lands.

But like many fugitive resources, including fish, water and air, hunted
wildlife resist enclosure, owing to their mobility, their intractability to
domestication, and their complex ideological associations with wild nature.
It is unsurprising, therefore, that at the dawn of the twenty-first century,
efforts to enclose wildlife are incomplete, with the promise of a fully privatized wildlife economy yet elusive. Even so, redoubled efforts to conclude
enclosure are ongoing throughout the U.S. West, with implications for the
relationships among state authority, capital, and non-human organisms.

This is not a phenomenon unique to the U.S. In Zimbabwe, 94 percent of
eland live on private game ranches, along with 64 percent of kudu, 63 percent of giraffe, and 56 percent of cheetah. Similarly in Namibia, since 1967,
when private wildlife ownership was instituted, private wildlife populations

have grown by 80 percent (Muir-Leresche and Nelson 2000), a fact reported and celebrated by neoliberal think tanks like the Competitive Enterprise Institute.

This study examines the efforts to privatize public wildlife in the U.S., pointing to both institutional and physical efforts at enclosure. Reviewing these efforts in the context of the changing political economy of the rural U.S. West, and with specific reference to the elk economy in the state of Montana, the study shows the way in which nature and labor resist efforts at enclosure. The combination of a rising incidence of Chronic Wasting Disease (CWD) and an increasingly well-organized anti-privatization movement points to general contradictions in capitalism's encounter with nature. A complex confluence of eco-managerial bureaucratic interests, gun populism, and virulent infection has created a barrier that is highly resistant to privatization efforts.

Rights to wildlife, conservation, and the game economy

The history of the western U.S. since the eighteenth century is one of enclosures. Native peoples were dispossessed of their lands through direct and violent means. These stolen resources, briefly in the public trust, were later turned over to private speculators in flagrantly corrupt land deals by the U.S. General Land Office. With the establishment of private rights in water, minerals, and forests, the remaining commons of the West marched through a steady pace of enclosures for 150 years, thanks to federal acts like the Preemption Act (1841), the Homestead Act (1862), and the Desert Land Act (1877) (Platt 1996).

State wildlife control

Despite this march of enclosure, some of the original commons of the U.S. West remain outside of the control of private capital. Most obviously, this includes wild animals. Legally, wild animals in the U.S. belong to the states in which they reside. This was established in the Constitution by default, insofar as all responsibilities not claimed by the federal government become those of the states. This was eventually clarified in legal precedent for wildlife through Justice Roger Taney's 1842 decision regarding fishing rights in Martin versus Waddell. Under this ruling, states hold the right to wildlife and individuals cannot claim private rights to wildlife simply because they hold land on which wild animals or fish are found. This decision was extended in the landmark case of Geer versus Connecticut, which established wildlife as part of the "public trust" – collective property of the people (Bean and Rowland 1997).

As a result, for the first 100 years of the republic, regulation and enforcement regarding wildlife were handled by the states, including the establishment of the first hunting limits, hunting seasons, and bans on commercial hunting, as well as the first deputized authorities and game wardens for enforcement. This authority hardened in the early twentieth century as states formed agencies and overseeing commissions to control and enforce laws and liaise with legislators forming policy (Leopold 1933).

Except where federal sanctuaries and reserves exists and when federal mangers and the strictures of the Endangered Species Act have come to take precedence, the states have reserved for themselves the right to manage significant wild species, especially those that historically have been hunted, many times to the brink of regional elimination.

State management, hunters, and collective interest

Historically the core of wildlife management is in the form of well-established state management regimes, therefore, which depend heavily on hunting and hunters as a management mechanism. Indeed, conservation of wildlife in the U.S. began with hunting-inspired regulation, growing into state efforts to control the decline not only of subsistence hunting animals but also the targets of sport and commercial hunting. State wildlife management policy and practice have co-evolved, therefore, with the demands and habits of hunting communities. As a result, hunting still represents the major population control for many major North American species.

A booming hunting economy has prospered over the last century that, although not without ecological problems, has maintained economically important species. When coordinated with wildlife reserves and the national park system, moreover, this anthropocentric and somewhat instrumental system of conservation has further helped to maintain non-economic species. State wildlife managers were early proponents of ecosystem-based approaches to environmental management, for example, and because they historically have had the trust of hunters, they have been able to implement bans and controls liberally in their pursuit of healthy and diverse ecosystems. Indeed, hunting constituencies tend to promote and defend these state level wildlife priorities, even when it means reduced take, enclosed areas, and year-to-year inconsistency in availability of game. The result has been the recovery of wild game and migratory birds in states where strong state wildlife management and hunting are integrated.

This hunting constituency is, however, bifurcated, with local hunters and hunting groups in tension with commercial hunters, commercial outfitters, and their out-of-state and relatively wealthy clientele. This split runs deep in the history of hunting, as a rift between the killing of wildlife as reproductive household practice and as productive industrial commerce. The earliest hunting and gun clubs were often formed to enforce state hunting season regulations, for example, against commercial hunters. Even Theodore Roosevelt's elitist Boone and Crockett Club, as early as 1887, worked to control and renounce commercial over-hunting (Zaslowsky and Watkins 1994). Though founded to further the private aims both of subsistence hunters and recreational sports hunters (two very different constituencies to be sure) local hunting advocacy compelled state controls, recognized collective ownership, and challenged commercial interests in the public domain, challenging the role of wildlife as an exclusive unit of production and profit with a vision of animals as collective elements of reproduction.

This division persists. Consider the differences in assets, priorities, and investments made by differing constituencies. In Montana, a typical case, the mean annual income of an in-state hunter falls between $30,000–35,000, while out-of-state hunters earn between $50,000–75,000. In-state hunters purchase gasoline and shells, spending $47 per day on average, while out-of-state hunters spend $207 per day without a guide and $478 per day with one (King and Brooks 2001). While 39 percent of in-state hunters rate meat procurement as a "very important" reason for hunting, only 17 percent of out-of-state hunters respond similarly (Allen 1988).

This hunting economy has further implications for the position of the state in the wildlife management. Historically, state wildlife agencies have depended heavily not only on the political clout of hunting constituencies to assure budgets, but also upon hunting licenses to fund ecological management. State wildlife agencies have a direct interest, therefore, in supporting the traditional management regime and the collectivist priorities that underpin it.

Economic transition, barriers to accumulation, and the emergence of enclosure

This contest over priorities occurs amid a larger political economic transition in all Western states. Ranching has been a dominant land use of the mountain West for the last century. Large parcels of grazing land and forest have remained under small producer control with the help of subsidies and stable markets for beef. With the emergence of feedlot-centered production systems, the vertical integration of the cattle industry, and meatpacking firms increasing monopsony power, producer margins have declined dramatically, putting traditional ranch properties in peril (Love and Burton 1999). As a result, there is an ongoing shift in land ownership from productive ranches to "amenity" ownership – where land in current production is purchased by wealthy out-of-state buyers who are interested in non-developed landscapes and good views. This transition is especially rapid around national parks and other "wildernesses" (Travis, Hobson *et al.* 2002). Rising land prices resulting from development further accelerate the sale of ranch properties and an out-migration from traditional communities. Thus, a production squeeze on primary production, especially ranching, coupled with a shift of new investment money into both recreation and development have set the terms under which the control of elk, mule deer, and other animals are contested today.

Collective control over these resources, while a boon for local hunters (and more generally for the populist culture of hunting), represents a barrier to accumulation for commercial interests in the West's changing political economy. For recreational outfitters, who sell hunting packages and represent a growing lobby in Western states, traditional state agency priorities – limited hunting seasons and licenses on the basis of ecosystem planning priorities – restrict avenues for profit taking. Further, non-transferable hunting licenses – which in-state hunters are guaranteed as a traditional right – bar the development of meaningful profits for competitive private game licensing. For ranchers facing tighter margins in a changing land market, the income stream from exclusive access to the wildlife that enters their property is attractive. Physical and institutional enclosures offer intriguing opportunities.

Institutional enclosure: "Ranching for Wildlife"

Enclosure of public wildlife for private good is barred by traditional state management regimes. This 150-year-old system is codified in laws in all 50 states. Successful enclosure, therefore, requires changes in the institutional structure of the distribution and transfer of access rights at the state level.

These "institutional enclosures," variously called "Wildlife Partnerships" and "Ranching for Wildlife," essentially follow the same model – rights to hunt are disseminated to landowners in large numbers (free of charge based on potential resident herd population), and the income value from sale and transfer are retained by the landowners and professional outfitting interests brokering the transaction. Use rights are retained by the individual hunter, but at highly differential costs (Leal and Grewell 1999).

Such rebundling of property rights so far has been enacted in eight Western states: California, Colorado, Nevada, New Mexico, Oklahoma, Oregon, Utah and Washington. In each case, owing to the statutory restructuring required to transfer rights from the state to the private sector, legislative action has been necessary. The geography of this transition, therefore, follows more generally the commitment of states to a larger agenda of privatization. The power of traditional hunting constituencies in some states (Idaho, Montana, and Wyoming) has proven a barrier to institutional change.

The character of formal enclosure programs vary in the eight states where they exist. Emerging since the early 1980s, all of them require hunting permits to be dispersed to landowners, and most require or allow extended hunting seasons relative to traditional state managed permits. The demands made on the landowner in return tend to vary. While some systems require some kind of management plan, habitat improvement, or public access, many do not, and simply transfer rights and income benefits directly from the state to private owners, largely based on acreage of holdings and habitat.

Economic theory of institutional enclosure

By placing game licenses in the hands of landowners and commercial guides with the right to transfer the hunt to the highest bidder, landowners are in theory provided with incentives for wildlife management and habitat improvement. Efforts in this direction are further touted as beneficial to the traditional, local hunting community, since they are supposed to ensure better breeding stock and healthier herds.

Spearheaded by so-called "New Resource Economists" at free market think tanks like the Property and Environment Research Center, institutionalized enclosures are touted as rational ways to avoid commons tragedies through the proper internalization of environmental externalities: the unremunerated expense and nuisance for land owners of wildlife management. These, like all problems, are best solved by markets. As its most vociferous promoter, Michael Copeland explains, New Resource Economists simply seek to "create private incentives and institutions wherever possible," with the hope that private ownership might replace government ownership (Copeland 1990: 23).

The notion that these efforts will result in overall improved ecosystem health is questionable. The incentive to produce high-value trophy animals provides no guarantee of ecosystem management and provision of biodiversity. So too, by reducing the authority of state wildlife management agencies (while still recruiting their efforts for emergency hunts, information provision, and other services at public expense), fee-based management decreases ecological control and flexibility, especially over the diverse geographies of migratory species. Enclosures do, however, shift the flow of value from public goods to private pockets, largely to the benefit of non-local elites, who in Texas have shown a willingness to pay up to $4,000 per animal, and to landowners who do not have statutory ownership of public animals (Leal and Grewell 1999).

Physical enclosure: game farming

While institutional efforts at enclosure are relatively recent and inchoate, physical efforts have a long history. The deer forests of nineteenth-century Scotland, as noted previously, represented private land that enclosed wild species. State law in the U.S., however, generally retards the development of such economies, since capture of wild animals represents a theft or unauthorized taking of state property.

Economic downturns in traditional agricultural sectors have, however, increased the incentive to produce and breed nontraditional game animals on land fenced for that purpose. As a result, there has been a marked expansion in game farms in the last several years, introduced to save small producers caught up in the rapid consolidation that has eliminated family farms throughout the country. Game farm operators offer a controlled fee-hunt and guarantee large trophy animals.

Like hunting, game farms are regulated largely by states, except where more general baseline regulations on animal welfare pertain. These fall under the jurisdiction of the Animal and Plant Health Inspection Service of the Department of Agriculture, a bureaucracy with a general interest in preserving and not over-regulating producers. As a result, game farms historically have been regulated using the same rules and enforcement mechanisms that apply to private zoos and animal exhibits, despite obvious differences between these types of businesses and hunting. Though rules vary from state to state, game ranchers cannot use captured animals, which are state property, so instead must either import animals from another state (and therefore from another game farm), or breed on site. Animals contained by fences require appropriate health certification and are subject to periodic inspection, quarantine, and systems of identification.

It is difficult to determine the number and total acreage of game farms in the U.S., both because there is no single federal register, and because the market is dynamic. In Montana in 2003, there were 77 operating game farm facilities, enclosing some 4,000 animals over 11,000 acres. This does not represent a large proportion of total animals or acreage, but it does seem to be an increasing trend; licenses for new and expanded facilities rose tenfold between 1993 and 1996.

Receipts from fee hunting on game farms and from sales of farm-bred wild animals are also unclear. While the USDA has advocated "alternative crops and enterprises for small farm diversification," game farms are one of several strategies (along with niche market commodities) that have done little to stem the decline of small family farms.

Resistance to enclosure

Like institutional efforts to award rights to wildlife to individuals, physical enclosures of animals represent another push towards the commodification of wild species. Fencing animals and reconfiguring the division of rights and responsibilities in nature to individuate the flow of value from biotic systems facilitates accumulation and fits neatly into neoliberal economic culture. This transition, though part of a steady movement first established in the eighteenth century, is not occurring without friction and resistance. Such resistance is inherent in the feedbacks that enclosures create.

Non-human consequences: confinement and Chronic Wasting Disease

CWD is a transmissible spongiform encephalopathy (TSE), a neurological disease in wild cervids (deer, elk, etc.) that produces small but fatal lesions in brains of infected animals. The disease is likely caused by abnormal infectious proteins without associated nucleic acids (prions). Other prion diseases are believed to include bovine spongiform encephalopathy (BSE, or mad cow disease), though no record of transmission between the two diseases has been demonstrated. Also, no direct connection has been established between CWD and Creutzfeld-Jacob disease, the form of TSE fatal to humans, although risks to those who hunt or consume deer or elk meat are increasingly being considered. The origins of the disease are unclear, but it has spread through wild deer and elk herds throughout the U.S. and Canada in the past two decades, resulting in massive infection rates among wild animal populations. While some states continue to seek confinement and control of the disease, many have resorted to mass slaughter of wild animal herds within quarantine zones (Cranmer and McChesney 2003; Salman 2003; Belay, Maddox *et al.* 2004).

Despite a lack of firm knowledge about the disease, one obvious pattern has emerged from recent observations: a key transmission vector is the interstate transfer of animals, specifically transactions between game farms. It is also likely that transmission rates are higher among captive herds. For both these reasons, the emergence and rapid spread of CWD have inspired increasing controls on game farm establishments.

Perhaps the most dramatic example is that of Montana, which largely eliminated the game farm industry in the wake of CWD outbreaks in other states during the mid-1990s. Ballot Initiative I-143, approved by Montana voters in 2000 and enacted as Montana Code 87-4-414 (2), states that: " ... the licensee (of a game farm) may not allow the shooting of game animals or alternative livestock, ... or of any exotic big game species for a fee or other remuneration on an alternative livestock facility (game farm)." This decision sounded the death knell for captive herds in the state, since in the absence of fee hunting, the only source of profit from game farming is the sale of animals between states. But because of the risks of spreading CWD through interstate animal sales, those transactions have plummeted. Prior to this, new licenses had been expanding rapidly, as landowners sought to diversify and capture benefits from commercialized wildlife (Montana Fish Wildlife and Parks Department 2003; see Figure 1.1).

Resistance from people: hunter ecopopulism

Resistance to market-based approaches continues to foment among local constituencies as well, who see in these efforts a pilfering of public goods. Hunters, anglers, and other users have begun to advocate against both game farming and restructuring of licensing rights, representing themselves through traditional "sportsmen" organizations.

In meetings of these organizations, the discourse surrounding the problem usually centers around the abstract idea of "access," a term that includes not only the right of entry onto public lands for hunting and fishing, but also the equal distribution of licenses and the right to cross private lands in order to access public property. This term of "access" is increasingly supplanted, however, with the more radical term – privatization. As ad hoc organizations like MADCOW (Montanans Against the Domestication and Commercialization of Wildlife) have contested legislation to deregulate game farms and restructure licensing, pro-access campaigns are increasingly conceptualized as anti-privatization (see Figure 1.2).

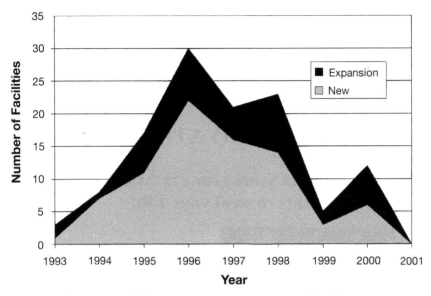

Figure 1.1 The rise and fall of game farming in Montana, 1993–2001.

This transformation is partly one of rhetoric, but it also reflects a shift in the broader consciousness surrounding the character of the public trust under changing demographic and economic conditions in the West. As flyers from the Montana Wildlife Federation (MWF) assert, "allowing commercialization of Montana's fish and wildlife will destroy sustainable, diverse populations in favor of market-driven production." Keeping wildlife as a public trust at the heart of their goals, MWF seeks

> [to] keep fish and wildlife in public ownership, allow no privatization or commercialization of wildlife or access to wildlife, make sure that fish and wildlife management remains in the hands of public agencies, and make certain the public's access to fish and wildlife is fair and equitable.

This resistance is reproduced and reflected in state resource management agencies. Wildlife managers remain relatively committed to collective and public solutions to game management questions and hostile to efforts to privatize wildlife. This is, in part, due to the benefits for agencies under current licensing arrangements, with receipts for licenses funding current enforcement activities as well as basic agency research. A shift in licensing institutions threatens the budgets of these agencies as well as their authority.

As a result, state agencies have attempted to broker access to private lands through the creation of state-funded subsidy programs. Block Management, a Montana state-sponsored program that pays landowners to manage land for wildlife habitat while allowing access to the public for hunting, is typical. The program is proffered as an alternative to enclosure and depends on state subsidy to create incentives against privatization.

More fundamentally, however, state wildlife agents articulate an ideological commitment to public trust approaches. First, game biologists and wardens insist that scientific management is impossible under individuated rights to wild species. Second, most connect the role of game management

WILDLIFE PRIVATIZATION

The Montana Stockgrowers Association wants to steal your Elk!

The Montana Stockgrowers Association, the same special interest group that has worked for years to keep you off your public lands, now has a plot to turn **public wildlife into private profit**. They call their plan the "Montana Wildlife Partnership", what it is is **Montana wildlife privatization.**

They want to profit from your elk, deer and antelope so they can get richer, and you get the shaft.

Don't wait until they sneak this scheme through the legislature, stop it now!

Ask your legislators how they will vote on this issue!

Demand an answer!

REMEMBER

ONLY YOU CAN PREVENT WILDLIFE THEFT!

Figure 1.2 Flyer from MADCOW.

not to concepts of "efficiency," an important discursive trope for market enthusiasts, but instead in normative terms of "equity." As one senior game official explained, "both doctors and ditch diggers should have a right to nature." These commitments extend agency activities to applying pressure with legislative allies to defeat institutional privatization reforms. In an unusual case of alliances between eco-managerial agencies and conservative populists, there are movements afoot to counter commercialized enclosure. Coupled with the disease feedbacks resulting from enclosed nature, barriers to privatization have emerged.

Eco-managerialism, gun-populism, and prions: barriers to enclosure

This outcome, an alliance between wildlife agency managers and conservative hunters is unlikely in several senses. First, as Luke has convincingly asserted, eco-managerial training and the agency culture it produces tend to direct the discursive and material practices of its managers in a manner subservient to the larger engine of capitalist modernity. "Well trained professionals, even when armed with sound science, can be used to serve the far more expedient goals of naked power agendas" (Luke 1999: 120). Fish and gun groups also may seem at first glance to be poor advocates for progressive and collectivist views of the economy. Native "Old West" communities tend to view nature in broadly utilitarian terms, to be politically conservative and/or libertarian, and to be vigorous defenders of private property rights (Reading *et al.* 1994). Finally, a disease ecology usually associated with domesticated and industrialized production systems also seems out of place in a discussion of wild animals.

The case of wildlife privatization suggests, however, a more complex reading of each of these socio-natural players. The soul of the "bloodsport" community is clearly split down the middle, across class and ideological lines, regarding its tolerance for hunting enclosures and the broader neoliberal agenda they represent. While many wealthy, non-local hunting advocates greet market-based solutions to wildlife "problems" enthusiastically, the prevalence of gun-populism in local communities has proven a remarkably heavy counterweight.

So too, the characterization of natural resource bureaucracy as "eco-managerial," while useful and persuasive, may overlook fundamental divisions within the state itself. While state agencies remain committed to semiautocratic and instrumental views of nature, their membership clearly shows more complex ideologies. Just as the environmental bureaucracies of the underdeveloped world – typically characterized as anti-popular, authoritarian, and environmentally ambivalent – have been shown to embody complex motivations, environmental commitments, and internal class struggles (Robbins 2000), state wildlife agencies in the U.S. embody contradictory commitments.

Finally, the emergence of CWD reflects a similar contradictory response to enclosure. Clearly, the unintended effects of physical enclosure for profit –

increased disease and reduced receipts from commercialized wildlife – represent a negative outcome of accumulation strategies in nature: realization of the underproduction (following O'Connor 1996) that emerges from the rising costs of reproducing the conditions of production. To argue that a "revolt" of nature has occurred simply because non-humans (elk, deer, etc.) have become commodified, however, is to under-specify the nature of the problem. More precisely, as with any effort to isolate or partition "economic" species in "economic" landscapes, away from "natural" species and "natural" environments, management invites the production of novel social/natural forms or "hybrids" (Latour 1993).

Crucially, then, this crisis is not simply inherent in neoliberalism; it is intrinsic to the modernity of contemporary management, rooted in an insistence on the ontological distinction between that which is natural from that which is social. This failed modern contract inevitably results in accelerated and pernicious evidence of its opposite, nature/culture hybrids like CWD. In this sense, the disease is best understood as a quasi-object, a hybrid fusion of human/non-human agents, propelled together precisely by efforts to pull them apart.

The emerging barriers to privatization and the neoliberal colonization of the non-human world, including progressive bureaucrats, radicalized hunters, and mutated prions, are catalyzed into existence by the very efforts of the privatization lobby. The failure of critical theorists and other observers is that of failing to notice these contradictory outcomes, soberly committed as they (we) are to the ontological categories inherited from static theories of neoliberalism. Whereas neoliberal environmental theorists are increasingly eager to use market solutions to help manage ecosystems, the unintended consequences and unlikely outcomes of this effort are somewhat harder to anticipate in oppositional theory.

While previous research has drawn attention to the potentially pernicious relationship among ecomanagerial bureaucracy, apolitical environmental activism, and specific consumption interests in nature, research cannot end there. Rather, it needs to seek out reservoirs of resistance, including and especially among non-progressive communities (e.g. hunters), and from within the walls of state technical bureaucracies (e.g. game wardens). So too, it must examine the specific ecological outcomes of institutional change, and rigorously trace their origins to their political economic roots. As we have learned internationally through ethnographic research of foresters and other technical managers, the practical politics of resource control open the door to unusual allies in the struggle over enclosure. This is nowhere as clear as it is in the case of wildlife – a final frontier for rapacious privatization.

Acknowledgments

This research was funded by a National Science Foundation Biocomplexity Grant (#BE-CNH #0216588). Thanks go to Julia Haggerty, Sheila McGinnis,

Dave McGinnis, and Dave Bennett. A version of this chapter first appeared in *Capitalism, Nature, Socialism* (2005), 16(1): 45–61.

References

Allen, S. (1988) *Montana Bioeconomics Study: Results of the Elk Hunter Preference Survey*, Helena, MT: Montana Department of Fish, Wildlife and Parks.

Bean, M. J. and Rowland, M. J. (1997) *The Evolution of National Wildlife Law*, Westport, CT: Praeger.

Belay, E. D., Maddox, R. A. *et al.* (2004) *Chronic Wasting Disease and Potential Transmission to Humans*, Atlanta, GA: Centers for Disease Control and Prevention.

Copeland, M. D. (1990) "The New Resource Economics," in J. A. Baden and D. Leal (eds) *The Yellowstone Primer: Land and Resource Management in the Greater Yellowstone Ecosystem*, San Francisco, CA: Pacific Research Institute for Public Policy.

Cranmer, M. and McChesney, T. (2003) "Chronic Wasting Disease: Risks to Hunters and Consumers of Deer and Elk Meat," *Neurotoxicology* 24(2): 73.

Dentry, E. (1999) "Buck Not Stopping Anywhere as Privatization of Wildlife Continues," *Rocky Mountain News*: 12C.

King, Z. and Brooks, R. (2001) *Montana Study and Results of the Elk Hunter Preference Survey*, Helena, MT: Montana Department of Fish, Wildlife, and Parks.

Latour, B. (1993) *We Have Never Been Modern*, Cambridge, MA: Harvard University Press.

Leal, D. R. and Grewell, J. B. (1999) *Hunting for Habitat: A Practical Guide to State–Landowner Partnerships*, Bozeman, MT: Political Economy Research Center.

Leopold, A. (1933) *Game Management*. New York: Charles Scribner's Sons.

Love, H. A. and Burton, D. M. (1999) "A Strategic Rationale for Captive Supplies," *Journal of Agricultural and Resource Economics* 24(1): 1–18.

Luke, T. W. (1999) "Eco-Managerialism: Environmental Studies as a Power/Knowledge Formation," in F. Fischer and M. A. Hajer (eds) *Living with Nature: Environmental Politics as Cultural Discourse*, Oxford: Oxford University Press, pp. 103–20.

Marx, K. (1967) *Capital*, vol. I. New York: International Publishers.

Montana Fish, Wildlife and Parks Department (2003) *Alternative Livestock Industry Fact Sheet*. Helena, MT: Montana Department of Fish, Wildlife, and Parks.

Muir-Leresche, K. and Nelson, R. H. (2000) *Private Property Rights to Wildlife: The Southern African Experiment*, International Centre for Economic Research.

O'Connor, J. (1996) "The Second Contradiction of Capitalism," in T. Benton *The Greening of Marxism*, New York: Guilford Press, pp. 197–221.

Platt, R. H. (1996) *Land Use and Society*, Washington, DC: Island Press.

Reading, R. P., Clark, T. W. *et al.* (1994) "Attitudes and Knowledge of People Living in the Greater Yellowstone Ecosystem," *Society and Natural Resources* 7(4): 349–65.

Robbins, P. (2000) "The Practical Politics of Knowing: State Environmental Knowledge and Local Political Economy," *Economic Geography* 76(2): 126–44.

Salman, M. D. (2003) "Chronic Wasting Disease in Deer and Elk: Scientific Facts and Findings," *Journal of Veterinary Medical Science* 65(7): 761–8.

Travis, W. R., Hobson, J. *et al.* (2002) *Ranchland Dynamics in the Greater Yellowstone Ecosystem*, Boulder, CO: Center of the American West, University of Colorado-Boulder.

Zaslowsky, D. and Watkins, T. H. (1994) *These American Lands*, Washington, DC: Island Press.

2 Privatizing conditions of production

Trade agreements as neoliberal environmental governance

James McCarthy

Trends in environmental governance

In this chapter and in the original, more detailed version in *Geoforum*, I argue that recent multilateral trade agreements, particularly the sweeping new protections they provide for investors, are redefining property rights and reshaping environmental governance in ways that are central to the neoliberalization of environmental governance. My central argument is that in addition to furthering the centuries-long process of the enclosure of nature under capitalism, the neoliberal agenda of NAFTA (the North American Free Trade Agreement) and similar trade agreements also involves something new: the privatization, or primitive accumulation, of conditions of production.

Ecological modernization theory, ecological economics, and a variety of other analytical approaches have asked what "globalization" and "neoliberalism" mean for environmental governance and quality. How to understand multilateral trade agreements is a critical topic in such conversations, one much debated in the rapidly growing literature on, "trade and the environment" (see Deere and Esty 2002). Those who believe that capitalist growth leads inexorably to improved environmental quality hail these agreements, arguing that the desire and ability to pay for environmental protection will follow in capital's wake. Conversely, many critics seem to take it as simply self-evident that trade liberalization is likely to undermine environmental regulation and quality, producing "races to the bottom." Ecological modernization theorists recognize the dangers of increasingly liberal trade, but still think trade agreements might be used to improve environmental governance (see Eckersley 2004). For instance, many saw NAFTA as an important step towards ecological modernization because, unlike previous trade agreements, it explicitly incorporated environmental considerations, and created new international institutions to address them (see Deere and Esty 2002).

In this chapter, I side with the skeptics, but try to present a more analytically precise argument regarding how and why these economically liberal investor protection provisions are likely to produce environmental degradation. Building on the work of Karl Polanyi (1944) and James O'Connor (1988,

1998), I argue that these provisions exemplify liberal capitalism's tendency to undermine the ecological conditions of its own reproduction. At their points of intersection with environmental regulation, these investor protections perform quintessentially liberal moves with respect to nature: they expand private property rights via enclosure, encourage the commodification of nature, and remove social constraints on capital accumulation. I argue that NAFTA's investor protection provisions both further the familiar enclosure of nature (nature-based *means* of production) and create new permutations of this process, by allowing privatization of nature-based *conditions* of production as an accumulation strategy. I suggest that the latter tactic can be usefully interpreted as *primitive accumulation of the conditions of production*. Finally, I argue that the forms of primitive accumulation analyzed here are cannibalistic on, rather than generative of, capitalist accumulation, and explicate why they are likely to accelerate the very crisis tendencies they seek to evade.

NAFTA and neoliberal trade

Since 1994, NAFTA has linked Canada, Mexico, and the United States in a regional trading bloc based on neoliberal principles of free trade. Negotiators still hope to extend its reach throughout the Americas via the FTAA (Free Trade Area of the Americas) (Deere and Esty 2002). On the other side of the Atlantic, the European Union has instituted sweeping liberalization within its borders, while other regions have formed trading blocs similar to NAFTA (e.g., Mercosur, the CACM, etc.). The World Trade Organization (WTO), created in 1994 to take over the functions of the Bretton Woods-era GATT, is by far the most extensive reorganizer of trade along neoliberal lines. And while a handful of high-profile, multilateral agreements receive the most attention, there are also over 1,800 bilateral trade agreements in existence, many of which share key features (Mann 2001).

These recent trade agreements, while varied, share common features that identify them as some of neoliberalism's major manifestations and mechanisms. First and foremost is their faith in "free markets" as institutions for allocating goods and resources, and specifically in exchange based on comparative advantage derived from factor endowments as the fundamental dynamic of economic activity and growth.[1] Their goals of increasing foreign direct investment and homogenizing regulatory environments across and within national territories follow from this position. Equally apparent in these trade agreements is a characteristically neoliberal antipathy and skepticism towards the state and its presumed inefficiency and unresponsiveness (Peck and Tickell 2002). Pursuing common neoliberal goals that follow from these foundations, NAFTA and other trade agreements seek to eliminate government subsidies to particular industries, shift important state functions to the private sector or to NGOs, devolve unavoidable state functions to the lowest level possible, and drastically reduce state spending (particularly in any area related to social welfare) (see Deere and Esty 2002 for examples).[2] These common neoliberal prescriptions can all be understood as contributions to an overarching goal of increasing the flexibility and profitability of capital (Peck and Tickell 2002, Tickell and Peck 1995).

Ideologically, these prescriptions share the conviction that governance is done best by actors other than states. Major multilateral institutions and actors, including the WTO, World Bank, United Nations, and many transnational corporations, have become increasingly aggressive in their willingness to look inside countries, evaluate their governance structures, and recommend both sweeping and highly specific changes. They focus not just on formal state operations but on more diffuse aspects of governance, such as the functioning of civil society, the robustness of institutions, and the quality and quantity of "social capital." Thus, one of the major trends in the liberalization of trade brought about via these agreements has been increasing penetration by trade authorities into the laws and regulations of individual countries (Mann and von Moltke 2002).

Such reconfigurations of governance along neoliberal lines encompass governance of the environment, as well. Although the trade agreements and specific provisions discussed here are not nominally about the environment, their effects on environmental quality and regulation might easily exceed those of explicitly environmental multilateral agreements. Trade agreements have already had major chilling effects on new environmental and public health regulations, as documented by the International Institute for Sustainable Development and the World Wildlife Fund (2001). Equally important is their potential to block many new interpretations or applications of existing environmental laws and regulations. Although NAFTA is sometimes lauded for including environmental provisions and indeed creating new environmental institutions (see Deere and Esty 2002), in practice these creations have done very little to remedy the environmental degradation resulting from the expanded market activity they help to legitimate (Sanchez 2002; Carlsen and Salazar 2002).

The ways in which capitalist interests trump environmental concerns in neoliberal practice can be seen by examining some of the trade agreement provisions most directly relevant to environmental governance. Two types of such provisions, the potential classification of environmental laws, regulations, and requirements as illegal "barriers to trade," and the substantial extension of protections for "intellectual property rights" to bits of nature within an increasingly globalized property regime, have received considerable attention and will not be discussed further here (see McCarthy 2004).

Investor protections

My central concern here is the substantial expansion of investor protections included in NAFTA. Identical investor protection provisions are included in current drafts of the FTAA, and very similar ones are to be found in WTO rules and many other trade agreements (see Deere and Esty 2002). Thus, it is fair to characterize these as widespread provisions of neoliberal trade agreements. Under these protections, foreign investors in a participating country are generally guaranteed the same treatment from the state as that received by domestic firms and by any of their non-domestic competitors operating in the country. They are protected against nationalization or other seizure of their assets, and cannot be required to have any domestic ownership, partners, suppliers, or markets. They are guaranteed that they will be treated in accord with international trade law and that they will be able to move investment funds and profits in and out of a country freely. The investor protection provisions attempt to specifically rule out strategies and tactics commonly employed by developing countries, many of them former colonies, in the decades following decolonization as they attempted to establish a more favorable position in the international economy and pursue national development projects. Common strategies and tactics included nationalization of foreign assets and industries established under colonial rule and strict performance requirements for new foreign direct investment, such as demanding that foreign firms have domestic partners, limit their repatriation of profits, transfer technology, and so on. When foreign firms contested such seizures and requirements, they often found themselves before unsympathetic or corrupt domestic judiciaries, and so the trade agreements also promise them different, international venues for dispute arbitration. In short, the investor protection provisions of recent multilateral trade agreements have been essential to securing the legal and political conditions for the dramatic transnationalization of production and finance in recent years, and to disallowing any lingering aspirations towards a developmentalist state.

NAFTA deals with investor protections in its Chapter 11. Its Articles 1102, 1103, 1105, and 1106 detail the investor protections discussed in the preceding paragraph. Article 1110 is of greatest interest here. Its purpose is to protect investors against direct expropriation, indirect expropriation, and measures "tantamount to expropriation." The first is concerned essentially with the government physically taking private property for purposes of state (the taking of a private house in the path of a new road is the standard example). Nationalization of entire industries is probably the strongest form of direct expropriation. "Indirect expropriation" and "measures tantamount to expropriation" are murkier; in fact, NAFTA includes them as categories of expropriation but does not clearly define them. Their core is the contention that any regulatory action by a government that reduces the maximum conceivable value of private property is a form of expropriation for which the government must pay the property owner. Such a doctrine dramatically expands both what counts as property, and what counts as expropriation.

The same argument circulated widely in the United States in the 1980s and early 1990s under the label, "regulatory takings." It was developed by

Richard Epstein, a conservative legal scholar at the University of Chicago (a hearth of neoliberal doctrine). Epstein contends that, "virtually any reduction in the use or value of private property due to regulatory action" constitutes a taking for which the government must pay compensation (Epstein 1985). Epstein's doctrine was adopted wholeheartedly by the Reagan administration in neoliberalism's first ascendant rush: Reagan's Executive Order # 12630 attempted to make the theory federal policy. It was thus in the realm of environmental governance that the U.S. first saw one of neoliberalism's signature strategies deployed: the effort to achieve *de facto* deregulation by making regulation prohibitively expensive, while calling for tax cuts and attacking budget deficits (Bromley 1991). Over the next decade, however, coordinated efforts to institutionalize this definition of regulatory takings throughout the United States, via court cases and ballot initiatives, largely failed. Just as the doctrine of regulatory takings was faltering in the domestic arena, though, extremely strong versions of it were being written into NAFTA and other multilateral trade agreements that had the potential to trump domestic environmental governance. This policy trajectory thus provides a concrete example of the neoliberal rescaling of governance to escape national-scale environmental constraints.

The idea of regulatory takings, or indirect expropriation, is inconsistent with most contemporary property theory. It assumes that completely unencumbered, asocial private property is the norm and that the state must pay its citizens for the exercise of its governance powers. Similarly, modern property law recognizes that regulation is allowed for legitimate public purposes. Regulatory takings arguments thus attack the notion of legitimate public goods. Likewise, most international law recognized until recently that states have "police powers," and that when they use those powers in nondiscriminatory ways to protect public goods (e.g., the environment), then expropriation has not occurred, even if private property has been affected (Mann and von Moltke 2002). Both domestic and international versions of the regulatory takings doctrine, then, are squarely part of the neoliberal project: they attempt to expand private property, shrink public goods and purposes, and roll back state regulatory powers and capacity. More broadly, they seek to use compelling discourses and narratives to restructure the culture of property in a new, neoliberal mold (see Rose 1994).

Dispute resolution

The dispute resolution procedures in NAFTA and other recent trade agreements also share important neoliberal elements. When a dispute is brought under NAFTA, the parties to the dispute can choose between two similar sets of rules for the arbitration: those of the World Bank's International Center for the Settlement of Investment Disputes (ICSID), or those of the United Nations Centre for International Trade Law (UNCITRAL). Both sets share the critical features below.

First, they are based on a model of corporate arbitration. Thus, market relations and norms structure the basic arena of disputes, rather than the domestic law of any given country, or international law dealing with subjects such as human rights or the environment. Second, investors – most often transnational corporations – are placed on equal footing with sovereign states in the proceedings, unlike most international law, which requires state-to-state interactions requiring states to act on behalf of corporations based in their territories. Corporations can initiate Chapter 11 proceedings against foreign governments without the approval or participation of their own governments; this is one of the features that distinguishes current trade agreements from their predecessors.

The cases are then heard, not by established judicial systems, but by tribunals appointed specifically for each case (usually from a list of international commercial arbitrators considered experts in international trade law). The investor bringing the complaint picks one arbitrator, the state being challenged picks another, and they agree together on a third. Companies initiating complaints presumably choose arbiters sympathetic to their positions (including extreme interpretations of expropriation provisions) (Mann and von Moltke 2002). Once a case has been initiated and tribunal members chosen, procedures specific to the case are agreed upon: tribunals are not bound by the legal procedures or norms of any particular country. Both written and oral arguments and evidence are usually presented, but the specifics are difficult or impossible to discover, because the proceedings are kept secret unless both parties to the dispute agree to make them public. Collectively, these changes represent a major redistribution of power between capital and states. Yet, as the following two cases illustrate, public laws and funds are at stake in investor protection proceedings.[3]

Cases

1 Methanex vs. the United States

Beginning in 1995, MTBE (methyl tertiary butyl ether), a gasoline additive and suspected carcinogen and neurotoxin, was discovered in groundwater in over 10,000 locations throughout California. It was presumed to have come from leaking underground gasoline tanks. In March, 1999, California banned the use of MTBE in the state, starting December 31, 2002. Many other states prepared to follow its lead.

The Canadian Methanex Corporation quickly challenged California's action by bringing a complaint against the United States under Chapter 11 of NAFTA. Methanex is a major producer of methanol, one of the ingredients of MTBE, and the company argued that California's action was an expropriation of its property. Specifically, Methanex claimed that future revenues that it would have earned in the absence of the ban had been effectively taken away from it, and it demanded $970 million in compensation.

It claimed that California had violated other of NAFTA's investor protection provisions, as well.

The Methanex case saw a crucial development in the operation of NAFTA tribunals. For the first time, environmental NGOs succeeded in gaining amicus (friend of the court) status in the case. NAFTA lays out many rules for tribunals and disputes, but says nothing one way or another about whether third parties can gain amicus status. Given that NAFTA adopts a narrow model of corporate arbitration, it seems unlikely that the drafters envisioned or would have welcomed such a possibility. Since the treaty does not preclude it, however, individual NAFTA tribunals appear to have the authority to accept or reject such involvement by third parties. The Methanex tribunal agreed to accept written submissions, but not testimony, from the environmental NGOs. This development replicated precisely one of the major pathways by which American environmental organizations first became major players in domestic environmental politics.

2 Metalclad vs. Mexico

In 1993, the Metalclad Corporation, a U.S. waste-disposal company, bought an inactive toxic waste dump from a Mexican waste management corporation. The dump site, in the municipality of Guadalcazar in the state of San Luis Potosi, was inactive because previous environmental and management problems had led to intense local opposition, centering on fears of contamination and allegations of increased cancer rates, and to federal orders that it be shut down. Metalclad sought to re-open the dump and expand it into a full-scale hazardous waste landfill. The company received some necessary permits from the federal and state governments. Metalclad later claimed that it had received verbal assurances that either no other permits were necessary, or that if they were, they would be forthcoming. The Mexican government denied having made such representations.

Metalclad began construction of an expanded facility without receiving any permit from the municipality of Guadalcazar, which ordered construction stopped. Metalclad applied for a municipal permit in November, 1994, and resumed construction without waiting for the result. Strong local opposition to the dump manifested itself in town meetings, organized protests, graffiti, and more: the prospect of a foreign firm ignoring local government and treating the reopening of the dump as if it were a purely private matter provoked intense anger. Guadalcazar denied Metalclad's request for a permit in December, 1995.

In 1996, Metalclad initiated an arbitration proceeding against Mexico under NAFTA's Chapter 11, claiming that Mexico had unclear, contradictory, and arbitrary standards at various levels of government about what permits were needed by investors and how to get them. It also claimed that its property had been expropriated by the municipal and state governments. In 1997, apparently in response to local sentiment and Metalclad's action under NAFTA, the governor of San Luis Potosi issued an "ecological decree" creating an ecological reserve for the protection of a rare cactus; the new reserve included the dump site. In theory, his action rendered the site ineligible for development regardless of the status of any permits. Moreover, the state government found that the dump would contaminate local water supplies.

A NAFTA tribunal found in Metalclad's favor in 2000 and ordered Mexico to pay the company $16.7 million in damages. The tribunal found that Mexico had "failed to provide a transparent, predictable framework for business planning and investment" and had expropriated Metalclad's property (Mann and von Moltke 2002).[4] Two aspects of the tribunal's ruling were especially troubling. First, it ruled that environmental regulation is a federal function in Mexico and that the municipality had therefore exceeded its jurisdiction by requiring a permit for the facility. Legal experts from Mexico's own government, however, had testified to the contrary. The tribunal's willingness to interpret and rule on domestic law, and to contradict domestic authorities in those interpretations, has grave implications. Second, the tribunal's ruling took no account of the state's purposes in the alleged "expropriation" of Metalclad's property. Any expropriation had been for the sake of protecting the environment and public health (Graham 2002), purposes long recognized in international law as legitimate uses of state power, even when private property is affected. The tribunal, however, said that the only relevant question when deciding whether expropriation had taken place was how much the value of Metalclad's property had been reduced by the state's action (Mann and von Moltke 2002), a criterion long advocated in "regulatory takings" circles.

Primitive accumulation of conditions of production

The investor protection provisions of NAFTA and other recent trade agreements contain both major continuities with classical liberalism's engagement with nature and a strikingly new twist on those perennial dynamics. The creation of new intellectual property rights in biological resources is best understood as the latest step in the centuries-long process of the enclosure of nature-based means of production under capitalism: the privatization of previously common or open access regimes by legislative fiat, backed up ultimately by the power of the state, and for the purposes of accelerating the commodification of nature. Such enclosures are a critical step in the process described by Marx (1967) as "primitive accumulation," meaning most centrally the separation of most producers from the means of production (De Angelis 1999).[5] They are technologically cutting edge, but theoretically quite familiar.

Yet I believe that NAFTA and similar trade agreements are also seeking something genuinely new. Discussions of primitive accumulation that center on environmental governance have focused on the privatization of nature as a means of production. Karl Polanyi (1944), James O'Connor (1988, 1998), and others have argued that conditions of production, including many environmental conditions, are different from means of production precisely because they cannot be privatized – due to their general nature, the impossibility of producing them as commodities, and the need for the state to monitor, regulate, and reproduce them on large scales. They have thus been framed as "impossible subjects" of enclosure. In fact, insistence on the protection of such conditions, whether theorized as the protectionist half of Polanyi's "double movement" or as O'Connor's socialization of the conditions of production, has often been understood as a direct reaction to the excesses of privatization and commodification and as an effort to re-suture the separations between populations and nature created by capitalism.

From this perspective, the rapid proliferation of environmental regulation in the United States from the mid-1960s through the 1970s can be viewed as an effort to establish common property in particular environmental goods at national scales. The same could be said of the growth of modern environmental movements in industrialized countries more generally. These movements pioneered new modes of democratic citizen involvement in postwar societies and had an enormous impact on environmental governance in formal and informal realms. This wave of regulatory action overlapped with high points in Keynesian-inspired policy in many realms, and with the tail end of the long Fordist expansion, when prosperity had become the norm and the realization of imminent crisis had not yet become widespread. Crucially, though, environmental strategies in this period focused largely on the national scale. Whatever the net effects of environmental regulation on economic competition at national scales (an ongoing debate beyond the scope of this chapter), strong national environmental laws and regulations did significantly constrain conditions of production. Corporations overwhelmingly resented anything that reduced their flexibility and accumulation strategies – particularly as the current era of more globalized competition began to unfold. A central goal of neoliberalism's environmental governance agenda, then, has been to free capital from these national-scale regulatory constraints (see McCarthy 2005). The trade agreement provisions discussed here relocate much environmental governance to international scales and into the hands of non-state judiciaries, and replace the openness in environmental governance created by the modern environmental movement with new forms of secrecy and closure.

I thus argue that what is new here is that trade agreement provisions such as those of Chapter 11, particularly those expanding the definition of "expropriation," are attempts to enact a primitive accumulation of the conditions of production. They attempt to guarantee stability in certain conditions of production as a brand-new private property right for specific firms, effectively privatizing what had been a public or common property right in those conditions. More simply: they create a property right centered not on enclosing nature in a way that keeps it from all others – the heart of the

liberal conception of privatization – but on privatizing the right to transform and exploit general, social nature in ways that will directly harm others.
Simply put, among the "property rights" being constructed here is the right
of some firms to pollute, the right to cause ecological harm and create
environmental hazards for people in a given area.

I believe this to be a defensible extension of the concept of primitive
accumulation for several reasons. One, because it is privatization: unlike
many previous efforts to resist environmental regulation, this tactic does not
seek to limit regulation of entire industries, substances, territories, and so
on – i.e., to contest general regulation of the conditions of production. Rather,
it seeks to secure differential conditions for specific firms by codifying them
as a new property right, saying that expropriation has occurred and compensation must be paid if a given firm's conditions of production change.
Two, and crucially, these privatizations do not emerge through the dynamics
of commodity production and circulation central to Marx's analysis of
capital accumulation, but rather are secured through legal maneuverings
and the mobilization of class power. In fact, they contradict the logic of
capitalism, and certainly the discursive self-representation of neoliberalism, in
key respects: corporations that sing the praises of free trade and comparative
advantage are simultaneously seeking to use "free trade" agreements to
move further away from a landscape of perfect competition by securing
differential conditions of production. Three, it is primitive accumulation of
the conditions of production, rather than of the means of production,
because it targets specifically the broad regulatory and police powers of
states over their own territories and populations.

Conclusion

Capitalist resistance to ever-more stringent and complex regulation of conditions of production is nothing new. But why and how did the specific
strategy of including sweeping investor protections in trade agreements
emerge as a leading form of such resistance over the past decade? Like
neoliberalism more broadly, these provisions are in large part a reaction to
rapid changes in international landscapes of competition since roughly the
early 1970s. Seeking to cut costs, increase flexibility, and maintain profitability, firms have paid ever more attention to the advantages available by
locating various portions of their production networks in competing
national territories. Since the end of the cold war in particular, firms have
come to take many of the basic functions of states more for granted, and so
have felt freer to choose among national territories from a truly global
menu. These decades of expanded and intensified competition have made
crises of accumulation appear far more likely.

The ultimate valorization of capital that firms have borrowed or advanced
through the normal circuits of capital – i.e., through the production of
commodities sold in competitive markets – is thus especially uncertain.

Locking in conditions of production for individual firms helps to ensure the ultimate valorization of their capital in two ways. First, the ability to secure exemptions from new requirements that may constrain their competitors is clearly a competitive advantage. Second and far more certain, though, "regulatory takings" claims allow firms to bypass the tremendously risky circuits of capital entirely, and to realize a healthy profit on their advanced capital by relying directly on the power of states to tax their populations. Rather than competing in the marketplace for years and negotiating all of the attendant risks, firms claiming projected future revenues as "expropriated" property simply demand cash, now, from national treasuries. This parasitic strategy falls squarely into the category of primitive accumulation.

Like most capitalist responses to looming crisis, these strategies set the stage for spatially and sectorally larger versions of the underlying problems down the road. States cannot long maintain environmental protections and hand out cash as "compensation" for hypothetical lost future revenues. Neither states nor markets could long survive the widespread adoption of such strategies, which are ultimately redistributive towards firms rather than true strategies for capitalist accumulation. Yet if states concede, effectively guaranteeing firms that their regulatory conditions of production will not change from the moment they make an investment, crisis also looms. It might be delayed as the environmental degradation in one place led to the relocation of production elsewhere, but the end result would be environmental degradation over large spatial scales, and thus the prospect of a larger crisis. Rejection of the "regulatory takings" premise is therefore essential.

Acknowledgments

My thanks to Jim Glassman, Scott Prudham, and anonymous *Geoforum* reviewers for extremely helpful comments on the original version of this piece, and to Elsevier for permission to reprint portions of the original article. Original article was McCarthy, J. (2004) "Privatizing Conditions of Production: Trade Agreements as Neoliberal Environmental Governance," *Geoforum* 35(3): 327–41.

Notes

1 Despite trenchant critiques showing that the assumptions underpinning comparative advantage are invalid (see Chang 2002).
2 These are general positions, of course; in practice, their application is far from consistent, with the most powerful countries, such as the U.S., often exempting themselves from many of the requirements they impose on weaker trading partners.
3 I examine these cases in far greater depth in the original article.
4 Mexico appealed the ruling to a Canadian court, the Supreme Court of British Columbia, under NAFTA provisions allowing for third-party review; this court upheld the tribunal's decision in 2001.

5 As an historical transition in England, primitive accumulation was an ensemble of processes that included the separation of laborers from the land, the conversion of multiple forms of property into the private property of capitalists, the creation of a wage labor force working at the edge of daily reproduction, slave labor, a deepening socialization of production, and the state's supporting use of violence and law (Harvey 2003). Debates continue over whether primitive accumulation is best understood as a purely historical process and category, one necessary to but obviated by the development of capitalism, or whether the dynamics it describes are ongoing both in space, as capitalism incorporates new areas and populations, and in time, as relationships still present in "mature" capitalist economies (De Angelis 1999; Perelman 2000; Harvey 2003). I discuss these issues and their implications for my argument in greater detail in the original article.

References

Bromley, D. (1991) *Environment and Economy: Property Rights and Public Policy*, Oxford: Blackwell.

Carlsen, L. and Salazar, H. (2002) "Limits to Cooperation: A Mexican Perspective on the NAFTA's Environmental Side Agreement and Institutions," in C. Deere and D. Esty (eds) *Greening the Americas: NAFTA's Lessons for Hemispheric Trade*, Cambridge, MA: The MIT Press, pp. 221–44.

Chang, H. (2002) *Kicking Away the Ladder: Development Strategy in Historical Perspective*, London: Anthem Press.

DeAngelis, M. (1999) "Marx's Theory of Primitive Accumulation: A Suggested Reinterpretation," available at: www.homepages.uel.ac.uk/M.DeAngelis/PRIMACCA.htm.

Deere, C. and Esty, D. (eds) (2002) *Greening the Americas: NAFTA's Lessons for Hemispheric Trade*, Cambridge, MA: The MIT Press.

Eckersley, P. (2004) *The Green State: Rethinking Democracy and Sovereignty*, Cambridge, MA: The MIT Press.

Epstein, R. (1985) *Takings: Private Property and the Power of Eminent Domain*, Cambridge, MA: Harvard University Press.

Graham, E. (2002) "Economic Issues Raised by Treatment of Takings under NAFTA Chapter 11," paper prepared for the 2002 ILSD Tri-National Policy Workshops: Mexico City, March 13; Ottawa, March 18, and Washington, DC, April 11.

Harvey, D. (2003) "The 'new imperialism': On Spatio-temporal Fixes and Accumulation by Dispossession," *The Socialist Register*.

International Institute for Sustainable Development, World Wildlife Fund (2001) *Private Rights, Public Problems: A Guide to NAFTA's Controversial Chapter on Investor Rights*, Winnipeg: International Institute for Sustainable Development.

Mann, H. (2001) "Case Notes," *Reciel* 10(2): 241–5.

Mann, H. and von Moltke, K. (2002) "Protecting Investor Rights and the Public Good: Assessing NAFTA's Chapter 11," background paper to the 2002 ILSD Tri-National Policy Workshops: Mexico City, March 13; Ottawa, March 18; and Washington, DC, April 11.

Marx, K. (1967) *Capital*, vol.1, New York: International Publishers.

McCarthy, J. (2004) "Privatizing Conditions of Production: Trade Agreements and Environmental Governance," *Geoforum* 35(3): 327–41.

—— (2005) "Scale, Sovereignty, and Strategy in Environmental Governance," *Antipode* 37(4): 731–53.

O'Connor, J. (1988) "Capitalism, Nature, Socialism: A Theoretical Introduction," *Capitalism, Nature, Socialism* 1: 11–38.

—— (1998) *Natural Causes: Essays in Ecological Marxism*, New York: Guilford Press.

Peck, J. and Tickell, A. (2002) "Neoliberalizing Space," *Antipode* 34(3): 380–404.

Perelman, M. (2000) *The Invention of Capitalism: Classical Political Economy and the Secret History of Primitive Accumulation*, Durham, NC: Duke University Press.

Polanyi, K. (1944) *The Great Transformation*, New York: Rinehart & Company, Inc.

Rose, C. (1994) *Property and Persuasion: Essays on the History, Theory, and Rhetoric of Ownership*, Boulder, CO: Westview Press.

Sanchez, R. (2002) "Governance, Trade, and the Environment in the Context of NAFTA," *The American Behavioral Scientist* 45(9): 1369–93.

Tickell, A. and Peck, J. (1995) "Social Regulation after Fordism: Regulation Theory, Neo-liberalism, and the Global-Local Nexus," *Economy and Society* 24(3): 357–86.

3 Dispossessing H₂O

The contested terrain of water privatization

Erik Swyngedouw

> Let us review the circumstances once more with special reference to the health of the workers ... they are deprived of all means of cleanliness, of water itself, since pipes are laid only when paid for, and the rivers so polluted that they are useless for such purposes.
>
> (F. Engels, *The Conditions of the Working Class in England*, 1845)

Accumulation by dispossession: privatizing H₂O

Dispossession or privatization?

'Making a buck' remains a tricky business, particularly under capitalism where predators, competitors, workers, hostile environments, and other assorted actors, tend to make life difficult for those noble spirits who endeavour to enterprise, seek out new ways of earning money, and invest scarce resources in the production of valuable things. Historically, capitalists have successfully managed to adapt and adjust to changing circumstances and conditions, eagerly exploring new avenues or returning to already tried ways of maintaining a healthy rate of accumulation. Over the past two decades or so, much of political-economic analysis has focused on how innovations and changes in both products and production processes constitute the backbone of competitive development and the necessary foundation to sustain both competitiveness and accumulation. In other words, political economy tended to concentrate on how capitalism functions best when capitalists do what they are supposed to do, i.e. to produce new things for the market under more efficient conditions of production. Of course, this process also kept the focus of attention on the capitalist labour process, and on the (re)organization of labour relations on the shop floor and class relations on the wider social level.

However, this classic form of accumulation through expanded reproduction (as Marx would call it) is of course only one of the possible avenues through which capital can expand and the accumulation rhythm be maintained. As David Harvey has observed in *The New Imperialism* (2003), alongside expanded reproduction, capital also expands by incorporating resources, peoples, activities, and lands that hitherto were managed, organized, produced under social relations other than capitalist ones. In fact, this 'primitive accumulation' or 'accumulation by dispossession' has become a key accumulation tactic in recent years. The official terminology for 'accumulation by dispossession' is of course 'privatization'. As the latter term suggests, privatization is a process through which activities, resources, and the like, which had not been formally privately owned, managed or organized, are taken away from whoever or whatever owned them before to a new property configuration that is based on some form of 'private' ownership or control. Privatization, therefore, is nothing else than a legally and institutionally condoned transfer of entitlements. Consider, for example, the wholesale sell-out of the state-owned or collectivized means of production and resources in former socialist countries, the successive waves of privatization around the world, and the international legal battles over property rights of gene plasma, water, indigenous knowledges, and so forth.

Of course, such tactics of accumulation by dispossession are embedded within a wider discursive and ideological frame that renders such policies not only legitimate, but normatively desirable. On the one hand, intellectual and theoretical arguments are advanced that signal the 'failure' of non-private modes of social organization of production. For example, state organization of production is these days invariably associated with failure. Indeed, 'state failure' in the delivery of services in developing production, in sustaining an innovative dynamics, in keeping economies competitive, has become the mantra of mainstream economists and of assorted conservatives and liberals. The second line of argument, in turn, revolves around the unequivocal celebration of market forces and private ownership. Mobilizing both moralistic arguments such as Hardin's Tragedy of the Commons and utopian arguments about the promises of development and success foretold by free market pundits, privately owned and market-organized production is invariably portrayed as leading to the most optimal output and the most socially desirable distribution of value. This twin argument forms the backbone of the current wave of neoliberalism that renders accumulation by dispossession as desirable as the next version of the Windows operating system.

Yet, 'making a buck' is indeed not that easy, even under neoliberal rule. For one, those dispossessed do not necessarily passively accept the theft of what they consider to be rightfully theirs. Secondly, once under the aegis of private capital accumulation, all manner of social tensions and conflicts arise. Predating competitors loom around the corner, recalcitrant workers raise the spectre of old and new forms of class struggle, disgruntled consumers mobilize the weapons of the weak when it becomes clear that the initial promises fail to materialize. And the state or other forms of collective institutional organization have to step in yet again to assure accumulation by dispossession keeps going notwithstanding the proliferation of social protests.

Over the last two decades, water has become one of the central testing grounds for the implementation of global and national neoliberal policies. The privatization of water production and delivery services, particularly urban water supply systems, has become an important arena in which global capitalist companies operate in search of economic growth and profits. The water sector, together with many others, has become one of the battlefields over which 'accumulation by dispossession' tactics are waged, often won by capital, and occasionally lost.

Tactics of dispossession: how did it happen?

Despite the raging debates over potential or actual shifts towards privatization, there is in fact a long history of changes in the urban water supply sector. Indeed, since the inception of urban water systems, they have always been characterized by shifting configurations of public–private partnerships and, consequently, by different types of property and control regimes. Most international studies demonstrate that the organization of urban water supply systems can be broadly divided into four stages (Hassan 1998). The first stage continued up to the second half of the nineteenth century, when most urban water supply systems consisted of relatively small private companies providing parts of the city (usually the richer parts) with water of varying quality (Corbin 1994). Water provision was socially highly stratified and water businesses were aimed at generating profits for the investors (Swyngedouw 2004). As Engels already contended, water pipes are laid where people can pay for the service; a process that is simultaneously excluding those deprived of the necessary means to access such privatized provision.

This was followed by a period of municipalization, primarily prompted by concerns over deteriorating environmental conditions and calls for a sanitized city (Gandy 2004). In the UK – as elsewhere in Europe – this took the form of a municipal socialism concerned with providing essential public goods at a basic, often highly subsidized, rate (Laski *et al.* 1935; Millward 1991). Profitability was without any doubt a secondary concern and subsidies came from the general tax income (from either the local or the national state).

The third phase started approximately after the First World War when the water industry, together with other major utility sectors (such as electricity and telecommunications), became part of a growing national concern. The national state, with varying degrees of intensity of control, regulation, and investment, took a much greater role in public services provision (Littlechild 1986). Water infrastructure became – together with other major infrastructure works and programmes – part of a Fordist-Keynesian State-led social and economic policy. The investments in grand infrastructure works (dams, canals, networks) were part of, on the one hand, an effort to generate and/or support economic growth, while, on the other hand, assuring a relative social peace by means of re-distributive policies (Moulaert and Swyngedouw 1987). Three objectives were central to this Fordist period of expansion of water provision: the creation of jobs, the generation of demand for investment goods from the private sector, and finally, providing basic collective production and consumption goods (like water, education, housing) at a subsidized price for wage workers and industry alike. In some instances, water provision was nationalized (as in, for example, the UK and many developing countries).

During the fourth and most recent phase, roughly starting with the global recession of the 1970s, a period associated with the demise of state-led economic growth and the subsequent transition to post-Fordist or flexible forms of economic development and state guidance (Moulaert and Swyngedouw 1987), a major shift took place in the public/private interplay in the water sector. First of all, mounting economic problems – in the context of high social and investment spending – resulted in growing budgetary difficulties for the national (and often also local) state. This necessitated a reconsideration of the direction of state spending and resulted in reduced expenditures in the welfare sector and in supporting debt-ridden industrial sectors or expansive infrastructure programmes (Ruys 1997).

Second, the call for greater competitiveness as a means to redress the economic crisis of the 1970s and early 1980s prompted a quest for efficiency gains and greater productivity through cutting red-tape, labour-market deregulation, and greater investment flexibility. This, in turn, was accompanied by privatization tendencies as a means to pursue both of the above recipe-solutions to the crisis of Fordism.

Third, the standard democratic, but corporatist, channels of government often infused by the presence and active lobbying power of social organizations – most notably unions – proved to be a considerable barrier to implementing swift policy-changes.

Fourth, the growing environmental problems and, consequently, the proliferating number of actual and potential conflicts in the management and regulation of the 'hydrosocial' cycle (Swyngedouw *et al.* 2002) proved to be a serious challenge for traditional forms of organization and implementation of water-related activities. Particularly in a context in which civil society-based environmental groups became more vocal and powerful, while access to new exploitable water reserves became more difficult, systems of governance had to become more sensitive to these issues. Particularly questions of restricting or controlling demand (demand management) as a strategy to lower water consumption and hence taking away the pressures on expanding the urban water resource base and ecological footprint of the city became more loudly heard. The internalization of all these tensions within a fundamentally state-owned and state-controlled sector like water became increasingly difficult (Swyngedouw 1998).

Finally, and perhaps most importantly, investors began to search for new frontiers for capital investment. Water presented itself as a possible new source to mobilize and harness as it offered the possibility for turning H_2O (again) into capital and profit. This privatization of the commons through a strategy of 'accumulation by dispossession' became increasingly central to accumulation dynamics as the standard routes of restructuring of existing capitalist-economic processes and investments in new products were no longer sufficient to absorb the ballooning volume of capital in search of profitable investment avenues. Indeed, water, together with other common pool goods like genetic codes, local knowledges, and the like, are rapidly becoming part of such accumulation strategies (Katz 1998; Bakker 1999).

Private H₂O, collective waters

In a context of commodification and demands for privatization, the traditional state-led way of managing the triad of demand–supply–investment decisions becomes fundamentally transformed. If the profit motive, either for public or private companies, becomes the yardstick against which performance is measured and the price signal a key instrument for regulating the demand/supply nexus, the contradictions between these moments in the economic process take a rather different turn. In an external context, in which expanding demand is seriously discouraged for environmental reasons, while investment needs to be maintained to extend, replace, and update the network, the balance sheet equations for water supply companies become rather specific. With a given demand structure, and increasing investment, profitability – and hence the sustainability of market-led water companies – can only be maintained via either productivity increases and/or price increases. But both are problematic. Productivity increases are generally capital and technology intensive and almost invariably lead to a rising organic composition of capital and a reduction in the work force. And while water rate hikes are possible, they remain politically sensitive and might lead to socially perverse effects.

With expansion of either total or per capita demand, the volume of profits can be maintained by means of an expansion of supply. In this context, it is interesting to note that the 'productivist' logic of water supply companies continues unabated – despite mounting calls for a more restricted water use. Furthermore, given the long-term and capital-intensive nature of investments in water infrastructure, there is a rather weak incentive to engage in major long-term and capital-intensive investment programmes. Put simply, there is a clear disincentive to invest in not directly profitable long-term activities like leakage control in contrast to productivity enhancing investments that improve short-term profitability. It is not a surprise, therefore, that the state or other parts of the public sector have to mediate these contradictions.

Dispossession and the state: a Faustian pact

Regulating dispossession

The water privatization business foregrounds also one of the central myths of the neoliberal model, i.e. that privatization means getting the state off the back of the economy and rolling back regulatory red tape. In contrast to this often-repeated refrain, in the water sector, the state or other governing arrangements are centrally involved in 'regulating' and 'organizing' privatization and dispossession. They change laws, rules, and conventions and produce new legal and institutional frameworks that permit and 'regulate' privatization, often imposing all manner of conditions that force privatization through. In addition, governments provide all manner of financial and other incentives to lure private companies, to foster private sector involvement, and the like. After privatization, a state-controlled regulatory institutional framework invariably has to be implemented, just to make sure that companies enjoying a 'natural' monopoly condition 'behave in competitive ways'.

New institutions, most notably in the field of economic and environmental regulation, accompany every privatization programme. As Bakker (1999) has pointed out in the context of the UK, the regulatory game that started with the privatization (and ostensibly deregulation) unleashed a certain 'regulatory creep', which has subsequently developed into a top-heavy institutional-regulatory body. Given the territorial monopoly-character of the privatized water companies, all sort of regulatory procedures, such as investment target-setting, pricing, environmental standards, abstraction and leakage standards, quality assurance, and the like, have been implemented.

The struggles over the boundary between the public and the private terrain operate primarily through two interrelated axes: first, environmental standards and, second, market imperatives. The tension between these becomes contained in the pursuit of environmentally friendly marketization, while the public/private tension is mediated through debates over the form that the commodification process should take. Unanticipated consequences of these debates are seen in the changing character of knowledge within the water sector. Information that was once in the public domain becomes commodified, takes on commercial significance and is often treated as confidential. In the context of a shift to governance, knowledge management is central to playing the regulatory game. Retaining control of technical institutions remains an important vehicle for government bodies (at a variety of scales) to preserve their relative advantage within negotiations.

In sum, rather than de-regulating the water sector, privatization has resulted in a profound re-regulation of the water market and in a considerable quasi-governmental regulatory structure. In the process, the set of social actors involved in the institutional and regulatory framework of the water sector has been significantly altered, with a new geometry of social power evolving as a consequence. This new choreography of institutional and regulatory organization is what we shall turn to next.

Dispossession and democracy: an unlikely twin

Needless to say, the transfer of water control and delivery from the public to the private sector involves a change in the choreographies of social and political power. With political and public involvement waning, the power of the citizen is reduced. Moreover, to the extent that water is turned into money and capital, and water users into water customers who pay for water (rather than being citizens entitled to access to water), the choreographies of political power around water are fundamentally overhauled. Principles of business secrecy, absence of participation, non-transparent decision-making procedures and the like characterize the privatized organization of the water sector. Although a vital and local good, the decision-making frameworks are taken away from local or regional political control and relegated to executive boardrooms of global companies.

A host of new institutional or regulatory bodies have been set up (in the UK appropriately called quangos (quasi-NGOs)) that have considerable decision-making powers, but operate in a shady political arena with little accountability and only limited forms of democratic control. These institutional changes have been invariably defined as part of a wider shift from government to governance (Swyngedouw 2000). Whereas in the past, water management and water policy were directly or indirectly under the control of a particular governmental scale – i.e., either at the national and/or the local (municipal) level – in recent years there has been a massive proliferation of new water-related institutions, bodies, and actors that are involved in policy-making and strategic planning at a variety of geographical scales.

The combined outcome of the above has been a more or less significant (very significant in the case of the UK, less so in the case of, say, the Netherlands) re-configuration of the scales of water governance. As Bob Jessop (1994) has pointed out for other domains of public life, the national scale has been re-defined (and partially hollowed out) in terms of its political power, while supra-national and sub-national institutions and forms of governance have become more important.

Privatization, in turn, has led to the externalization of a series of command and control functions. The result is a new scalar 'gestalt' of governance, characterized by a multi-scaled articulation of institutions and actors with varying degrees of power and authority. The overall result, therefore, is a 'glocalization' of the national government, both upwards to the supra-national level and downwards to the sub-national level (Swyngedouw 1989; 1997). This results in a more complex articulation of varying geometries of scale-dependent forms of governance. In sum, national governmental regulation is simultaneously up-scaled and down-scaled, with an accompanying change in the choreographies of power, both between and within institutions.

Finally, accumulation by dispossession itself, of course, results in much greater power and autonomy for the companies themselves in terms of strategic and investment decisions. Privatization *de facto* means taking away some control from the public sector and transferring this to the private sector. This not only changes decision-making procedures and strategic developments, but also affects less tangible elements such as access to information and data. Traditional channels of democratic accountability are hereby cut, curtailed, or re-defined. A plethora of new institutions has been formed at a variety of geographical scales. This proliferation of 'governing bodies' has diminished the transparency of the decision-making process and renders it more difficult to disentangle and articulate the power geometries that shape decision-making outcomes. In practice, it can be argued that the transition from government to governance has implied – despite the multi-plication of actors and institutions involved in water management – the transfer of key economic and political powers to the private component of the hydrosocial governance complex. This, however, has not happened in a social vacuum and has rather fuelled a constellation of social and political conflicts, not least because of the consequences of an increasingly private-oriented governance model for the sustainability of socio-environmental systems.

Cracks in the mirror: the contradictions of dispossession

The supply/demand nexus and the investment/pricing conundrum

At a moment when the price signal becomes a central organizing principle of water markets, and in a context of relatively fixed supplies, demand management becomes tricky business. Monopolistic market control that is inevitably associated with water supply networks demands a strong price-regulation by the state or other governmental agencies. In addition, efforts to reduce water consumption for environmental reasons are countered by cost-recovery requirements that hinge on price setting and produced quan-tities. Invariably, water companies are operating in the two-pronged wedge of price-setting regulatory systems on the one hand and costly technologi-cal/organizational investments to enhance productivity on the other. The triad of investment/price/supply becomes very difficult to manage, particu-larly in a context of increasing pressures to reduce demand. There is a continuing tendency to increase supply despite rhetorical attention to demand management. The costly introduction of water-saving technologies is, at best, slow, while major efforts are made to increase supply despite often formidable opposition. It is becoming abundantly clear that the price signal is insufficient to regulate the allocation and efficient use of a resource like water.

Globalization through shared control

At a global scale, an accelerated process of concentration and consolidation is taking place that is rapidly leading to a fairly oligopolistic economic structure of water utility companies, with two (French) companies control-ling about 70 per cent of the global privatized water market (Hall 1999). This tendency has been further accentuated by the recent collapse of Enron, one of the leading global multi-utility companies. This raises difficulties of regulating global companies (particularly with respect to environmental and social standards, investments, maintenance and infrastructure upkeep). Indeed, the 'market' does not exist as a playing field without the actors making it work. The small number of global water companies produces an oligopolistic form of market organization. Only a handful of companies control the water market. In fact, two French companies, Ondeo (Suez) and Vivendi (now Veolia), take an overwhelming share of the water market, with Thames Water and SAUR trailing far behind in respectively third and fourth place. The dominance of the French is related to their long-term preferential access to the French water market. This gave them a competitive edge in international markets once they became more deregulated and were prepared for the privatization onslaught. Moreover, the French had more than a century of experience with a public–private partnership model in which the public sector owned the infrastructure while the private companies managed the water service. The Anglo-Saxon model is rather based on full privatization (infrastructure and delivery) and the export of this model has resulted in several failures or under-performing utilities.

Cherry-picking as strategic device

Servicing urban residents with reliable potable water services is not an easy business. It requires significant long-term investment, and complex organizational and management arrangements. And profitability is by no means assured, particularly in urban environments where many people have a low ability to pay and problematic access conditions (Swyngedouw 2004; Heynen *et al.* 2006). In short, only some urban water systems are likely to generate the prospect for long-term profitability, while others will continue to require subsidies and support if they are to continue to improve service delivery. Recent experiences have indeed shown that global private companies only really go for the nice bits; those that have some meat on the bone. That means that only big city water works are considered worthy of privatization.

Corruption as institutionalized practice

The inevitably strong link between the state and the private sector in privatization schemes opens up all manner of corrupt practices. They may be illegal, but more often than not, belong to the standard arsenal of agreed practices and accepted procedures. Needless to say, forms of bribery, under-the-table deals, greasing hands to facilitate certain contractual arrangements and financial contributions to political allies, all belong to the standard tool-kit of privatized water utilities. The concession contract for Jakarta with Thames Water had to be renegotiated after allegations of corruption. Bribery scandals were also associated with the concessions in, among others, Grenoble, Tallinn, Lesotho and in Kazakhstan. Enron, Vivendi, and Suez have all been accused of making payments to political parties in return for favours. Equally unsubtle but fully legal inducements for privatization are offered by national states and international organizations.

Water and market risk: the globalization of water and uneven development

To the extent that water companies operate increasingly as private economic actors, they are also increasingly subject to standard market risks. While providing a fundamental and essential service, the economic survival of water operations is not necessarily guaranteed. Takeovers, disinvestments, geographical re-allocation, bankruptcies, inefficient operations, political risk, and the like are endemic to a private market economy. In fact, this uncertainty and fluidity are exactly what market dynamics are supposed to produce, i.e. to weed out under-performing companies, and to re-allocate economic resources from less to more profitable activities. This raises particular questions with respect to the long-term sustainability of market-based urban water supply systems. In the absence of strong incentives to enhance productivity or efficiency, and given the high cost and long time horizon of fixed capital investments in water infrastructure, private companies may fail to keep water systems running efficiently.

Struggling against dispossession: a chequered success

Needless to say, the processes outlined above do not go uncontested (Petrella 1998). A plethora of local and global resistance movements have sprung up that contest the hegemonic logic of water privatization and fight for alternative modes of water management (The Center for Public Integrity 2003; Corporate Europe Observatory 2005; Sjölander Holland 2005). The twin tensions between continuing increasing demand for urban water on the one hand and the mounting pressure to allocate water to other functions on the other have proliferated socio-spatial tensions and conflict over water abstraction, water allocation, and water use.

Accumulation by dispossession in the water sector has indeed generated all manner of social struggles and conflict that highlight the fundamental antagonisms that are generated through such tactics. Of course, the plethora of social struggles is as broad as the range of strategies and tactics of dispossession. While social struggles within the dynamics of expanded reproduction of capital generally take a clear class-character, the social struggles around tactics of dispossession show a rather different choreography, usually centred on questions of ownership, control, participation, community interests, and the like. While all these forms of struggle are related to and generated by the dynamics of accumulation, the specific choreography of such struggles takes specific and often highly localized forms. This complicates attempts to form trans-local coalitions because both the stakes and conflict arena may be vastly different from place to place. Yet, the mobilization against strategies of accumulation by dispossession in the water sector is widespread.

Myriad struggles not only testify to the deep discontent communities express when confronted with pervasive tactics of dispossession, but they also signal to the powers that be that such blatant neoliberal class politics will face considerable resistance. This in itself will make the chances for successful privatization – i.e. turning local waters into global capital – less of a certain prospect. Indeed, increasingly, water companies themselves find that the promised honey-pots of large profits in the water business may not be as plentiful as portrayed by the World Bank and other pundits of liberalization. Some have begun to withdraw from the water sector. Water does indeed remain a highly contested good. And in a context in which still far too many people die from lack of access to good quality water, the social struggle for water has to be turned into a struggle for fundamental human rights.

Acknowledgments

A longer version of this chapter was Swyngedouw, E. (2005) 'Dispossessing H₂O: The Contested Terrain of Water Privatization', *Capitalism, Nature, Socialism* 16(1): 81–98.

References

Bakker, K. (1999) 'Privatizing Water: The Political Ecology of Water in England and Wales', unpublished DPhil thesis, School of Geography and the Environment, University of Oxford.

Corbin, A. (1994) *The Foul and the Fragrant,* London: Picador.

Corporate Europe Observatory (2005) *Reclaiming Public Water*, Amsterdam: Transnational Institute.

Gandy, M. (2004) 'Water, Modernity and Emancipatory Urbanism', in L. Lees (ed.) *The Emancipatory City: Paradoxes and Possibilities,* London: Sage, pp. 178–91.

Hall, D. (1999) 'The Water Multinationals', Occasional Paper (mimeographed), Public Services International Research Unit, University of Greenwich.

Harvey, D. (2003) *The New Imperialism*, Oxford: Oxford University Press.

Hassan, J. (1998) *A History of Water in Modern England and Wales*, Manchester: University of Manchester Press.

Heynen, N., Kaika, M. and Sywngedouw, E. (2006) *In the Nature of Cities: Urban Political Ecology and the Politics of Urban Metabolism*, London: Routledge.

Jessop, B. (1994) 'The Transition to PostFordism and the Schumpeterian Workfare State', in R. Burrows and B. Loader (eds) *Towards a PostFordist Welfare State?* London: Routledge, pp. 13–37.

Katz, C. (1998) 'Whose Nature, Whose Culture? Private Productions of Space and the "Preservation" of Nature', in B. Braun and N. Castree (eds) *Remaking Reality: Nature at the Millennium*, London: Routledge, pp. 43–63.

Laski, H., Jennings, J., Ivor, W. and Robson, W. A. (eds) (1935) *A Century of Municipal Progress 1835–1935*, London: George Allen & Unwin

Littlechild, S. (1986) *Economic Regulation of Privatized Water Authorities*, London: HMSO. Parker, D. (1997) 'Privatization and Regulation: Some Comments on the UK Experience', Occasional Paper No. 5, London: CRI, CIPFA.

Millward, R. (1991) 'Emergence of Gas and Water Monopolies in Nineteenth Century Britain: Contested Markets and Public Control', in J. Foreman-Peck (ed.) *New Perspectives in Late Victorian Economy: Essays in Quantitative Economic History 1860–1914*, Cambridge: Cambridge University Press.

Moulaert, F. and Swyngedouw, E. (1987) 'A Regulation Approach to the Geography of the Flexible Production System', *Environment and Planning D: Society and Space* 7: 327–45.

Petrella, R. (1998) *La Manifeste de l'eau*, Brussels: Editions Labor.

Ruys, P. (1997) 'Structural Changes and General Interest: Which Paradigms for the Public, Social and Cooperative Economy?' *Annales de l'économie publique sociale et cooperative* 68: 435–51.

Sjölander, H. (2005) *The Water Business: Corporations Versus People*, London: Zed Books.

Swyngedouw, E. (1989) 'The Heart of the Place: The Resurrection of Locality in an Age of Hyperspace', *Geographiska Annaler B* 71B(1): 31–42.

—— (1997) 'Neither Global Nor Local: "Glocalization" and the Politics of Scale', in K. Cox (ed.) *Spaces of Globalization: Reasserting the Power of the Local*, London: Longman, pp. 137–66.

—— (1998) 'Homing In and Spacing Out: Re-Configuring Scale', in H. Gebhardt, H. Günter and R. Wiessner (eds) *Europa im Globalisierungsprozess von Wirtschaft und Gesellschaft*, Stuttgart: Franz Steiner Verlag, pp. 81–100.

—— (2000) 'Authoritarian Governance, Power and the Politics of Rescaling', *Environment and Planning D: Society and Space* 18: 63–76.

—— (2004) *Social Power and the Urbanisation of Water: Flows of Power*, Oxford: Oxford University Press.

Swyngedouw, E., Kaika, M. and Castro, E. (2002) 'Urban Water: A Political-Ecology Perspective', *Built Environment*, 28(2), 124–37.

The Center for Public Integrity (2003) *The Water Barons*, Washington, DC: Public Integrity Books.

4 Neoliberalism in the oceans

"Rationalization," property rights, and the commons question

Becky Mansfield

Putting property at the center

To the extent that neoliberalism, with its calls for letting "the market" address myriad social and economic woes, has become the dominant model for political economic practice today, it should be expected that environmental governance, too, would be shaped by the neoliberal imperative to deregulate, liberalize trade and investment, marketize, and privatize. Here, I analyze development of neoliberalism in the oceans, and in particular in ocean fisheries. Examining ways that past policy orientations toward fisheries have influenced development of neoliberal approaches to ocean governance, I contend that neoliberalism in the oceans centers specifically around concerns about property and the use of privatization to create markets for governing access to and use of ocean resources. Within the Euro-American tradition that has shaped international law of the sea, oceans were long treated as common property, open to all comers with the means to create and exploit oceanic opportunities. Recent decades, however, have seen a pronounced shift away from freedom of the seas. Responding to new economic desires, environmental contradictions, and conflict over ocean resources, representatives from academia, politics, and business increasingly call for enclosing the oceans within carefully delimited regimes of property rights, be those regimes of state, individual, or collective control.

At the center of this new political economy of oceans, as it has evolved over more than 50 years, has been concern about "the commons," and the extent to which common and open access property regimes contribute to economic and environmental crises, which include overfishing and over-capitalization. As such, the question of the commons has been at the center of numerous, seemingly contradictory approaches to ocean governance and fisheries regulation. Thus, the first argument of the chapter is that neoliberal approaches in fisheries cannot be treated simply as derivative of a larger neoliberal movement that became entrenched starting in the 1980s. Instead, examining trajectories of neoliberalism in fisheries over the past half-century reveals that the emphasis on property and the commons has contributed to a more specific dynamic of neoliberalism operating in ocean fisheries and, therefore, to distinctive forms of neoliberalism.

A second argument is that focusing on property regimes as a key factor in the political economy of fisheries has contributed to convergence of quite different approaches around neoliberal, market-oriented perspectives. These approaches include not only neoclassical fisheries economics, but also approaches that focus on extended state jurisdiction over the oceans and on community management of fish resources. Despite their differences, many proponents of these viewpoints take property as their central problematic and contribute to the idea that proper specification of property rights is the foundation on which proper use of ocean resources rests. When scholars and managers make a case for property rights in fisheries as an alternative to open access regimes, they are not simply arguing for the importance of environmental governance in general. Instead, they make the neoliberal argument that property rights can harness people's supposedly innate profit motives for the good of all.

A third argument, running implicitly throughout the chapter, is the importance of the state in neoliberal regulation. Whereas proponents offer neoliberalism as an alternative to state governance, this chapter shows that privatization, in particular, relies on states to create and maintain property rights. Whether it is in the form of enclosing the oceans as state property, deciding how to further devolve property rights to individuals and collectives, or enforcing those property rights, states have been central to the neoliberal shift in ocean governance.

The chapter first discusses development of the idea of property rights in fisheries, primarily though not exclusively in the US. It starts in the 1950s–1960s, when neoclassical economic analysis entered into fisheries, and then traces the threads of property rights and the commons through the move toward extended political jurisdiction over the oceans in the 1970s and 1980s and the emphasis on common property management from the 1980s to today. By the turn of the millennium, these different approaches to the governance of fisheries all converged around the neoliberal notion of creating market incentives by specifying as property the right to fish. The chapter then illustrates these shifts by describing the suite of recent changes in the fisheries of the US portion of the North Pacific. Fishery regulators have pioneered distinctive modes of privatization that incorporate collective as well as individual forms of private property. The fisheries of the North Pacific are one example of how emphasis on the commons is at the heart of neoliberal privatization of the oceans.

The commons question

Fisheries, economists and the common property dilemma

In 1954, H. Scott Gordon wrote an influential paper about the fishermen's problem in which he argued that economic inefficiencies and overexploitation are inevitable in fisheries as long as fish are treated as a common, rather than private, resource (Gordon 1954). Predating the more general "tragedy of the commons" argument by over a decade (Hardin 1968), Gordon's main argument was that the lack of private property drives a non-equilibrium pattern in fisheries. Without private property "there is no assurance that [the fish] will be there for [the fisherman] tomorrow if they are left behind today," because someone else might catch them today (Gordon 1954: 135). Therefore, the rational fisher will catch as many fish today as possible. According to this neoclassical economic model, this leads inevitably to overcapitalization, as, first, each fisher applies more capital (e.g. new technology or more hours at sea) to extract fish before everyone else, and, second, as new fishers enter the fishery to capture rents. These two forms of expansion then lead inevitably to rent dissipation, as overcapitalization absorbs any potential profits, and to overexploitation, as the race for fish requires that each fisher catch as many fish today as possible. This then drives increasing capitalization and effort as fishers compete for ever diminishing fish and profits, and so on. In this view, it is impossible to have an efficient and environmentally friendly industry for a common property resource.

Gordon's paper marked a historic moment in the development of fisheries economics (Scheiber and Carr 1998; St. Martin 2005). The paper argued that fisheries management is as much or more about the economic actions and decisions of people as it is about fish themselves, and thus fishers (and especially their economic decision-making) should be seen as endogenous to fisheries systems. The solution to problems in fisheries, then, is not to focus on fish and their biological condition, but is to focus on economic efficiency and ways to reform the property regime to harness individual decision-making to both market and ecological realities. Following this, throughout the 1950s and 1960s economists (particularly in the US) continued to develop their ideas about the problem with common property, including the fallacy of creating economic *inefficiencies*, such as limitations on gear, as solutions to fisheries problems. Their alternative was to develop (in theory, not yet in practice) mechanisms for "limiting entry" to individual fisheries as a way of moving toward property rights. Therefore, at the center of the emerging and increasingly influential field of fisheries economics was the commons, which was cast as a market failure: in the absence of clearly specified property rights, rational individual behavior leads to economic and environmental problems. Economic efficiency then becomes *the* key to social and environmental welfare: individual rational decision-making in free markets results in the greater good for all. Although these neoclassical fisheries economists of the 1950s–1960s did not completely dismiss state involvement in fisheries, they do prefigure by several decades the *laissez-faire*, free market themes of contemporary neoliberalism (which builds on neoclassical theory to create a larger political project). In this way, current neoliberal themes of market mechanisms and the importance of privatization are central to the birth of social science approaches to fisheries and concerns about control over and access to the oceans.

Enclosing the global ocean commons

During this same period the ocean regime shifted from one of primarily open access to one in which individual coastal states had sovereign rights to control and exploit economic resources in large areas of ocean adjacent to their land (Steinberg 2001). Individual states, first unilaterally and then under international law, extended their political economic jurisdiction generally from three nautical miles from shore to 200 nautical miles. The process started in the 1950s when several Latin American countries declared 200-mile zones in an effort to protect "their" resources from distant water fishing fleets from the global North. This sparked three decades of UN conferences on the Law of the Sea, and by the time these 200-mile zones became customary international law in the early 1980s, approximately 30 percent of the world's oceans (Nadelson 1992) and 95 percent of the world's fish catch (Juda 1991) were enclosed as state property.

This form of limited access seems to directly contradict neoliberal approaches to markets and states, in that political enclosure represents an expansion, rather than a limitation, of state control and governance. Yet, extended jurisdiction in many ways is consistent with the economic argument about property rights and economic efficiency, and fisheries economists generally supported extended political jurisdiction as a form of property rights (Caddy and Cochrane 2001; Christy and Scott 1965; Scheiber and Carr 1998). First, extended jurisdiction encloses the global commons as state territory, creating a new form of property right. Second, it provides the foundation from which states can further enclose the oceans through limited licenses or other privatization schemes. By enclosing oceans as national territory, extended jurisdiction is the first step toward further devolving property rights to individuals. In reality, however, much of the fisheries management effort after extended jurisdiction went toward expanding domestic fisheries effort; state programs often helped finance vessel construction and develop new domestic and international markets (Mansfield 2001a, 2001b). Seeing this, many economists began to argue that the economic incentives of state management were in fact irrational (Crutchfield 1986). In this view, blame for problems is placed not on specific regulations and programs, but on the property regime itself: state property is treated as a form of open access. Further, even when states did use their new management authority to implement license limitation programs, fishers (often with state support) were able to continue to capitalize by redesigning vessels and other technology: effort can increase even if the number of vessels stays the same.

Thus, this period, in the 1980s, marks the point at which fisheries economists adopted a fully neoliberal approach that disregards state involvement in fisheries (e.g. Neher *et al.* 1989). This move was not primarily due to the more general political shift toward neoliberalism at this time, but was influenced by the realities of fisheries development in the era of extended jurisdiction as viewed through the lens of several decades of economic theory on the commons and property rights. Economists used extended jurisdiction as an object lesson in how open access to a common property resource leads to inefficiency and degradation, and they continued to advocate for privatization of fisheries.

Institutions and rules: the commons as a form of property

By the 1980s, scholars in anthropology, institutional economics, and geography, among other fields, were challenging this "tragedy of the commons" model that posits the commons as the ultimate cause of environmental and economic problems associated with resource use. Researchers found numerous case studies from around the world in which local people successfully managed common property resources using combinations of explicit and implicit rules and cultural norms, and from this empirical starting point offered the commons not as the underlying cause of resource problems, but instead as a potential solution (Berkes *et al*. 1989). Scholars have then used this new "benefits of the commons" model in at least two ways. One shifts the focus away from property dynamics *per se* (i.e. if the commons is not the problem, then what is?), while the other stitches together empirical insights on the benefits of the commons and theoretical focus on property rights. I focus here on this latter strand of the literature, which centers on developing a more precise variant of the "tragedy" model, retaining the focus on specification of property rights as the solution to resource problems.

Scholars in this area work to make explicit some of the assumptions of orthodox commons models, and then evaluate those assumptions using empirical evidence as well as abstract logic. They then modify those assumptions deemed faulty, with the aim of making the model more robust and more useful for devising solutions to resource problems (Berkes *et al*. 1989; Ostrom *et al*. 1999). The first assumption that scholars challenge is that common property and open access are the same. In contrast, researchers define common property as that which is owned and controlled collectively, and distinguish it from that which is not owned and controlled by anyone (open access). The second assumption they challenge is that institutions (including rules and social pressures) play no role in constraining individual's actions. In contrast, researchers specify that what makes the commons successful are precisely the social institutions and cooperative agreements that are posited as insignificant, or impossible, in the "tragedy of the commons" model; instead, economic rationality, in which profit maximization is the driving force behind individual decision-making, can be modified by social practices. In this view, then, common property belongs in the category of possible types of property rights, along with private and state property, rather than in the category of open-access non-rights.

My argument is that this strand of the commons literature does *not* reject the underlying economic approach that defines the commons as the problem (see also Johnson 2004). Instead, these scholars specify the commons model by more carefully articulating types of property and their relations to social institutions. At the same time that they carefully articulate the complex social dynamics of common property, these scholars transfer the idea of the "tragedy of the commons" to that of the "tragedy of open access," such that the problems often associated with commons are attributed instead to open access (Mansfield 2001a). One commons proponent stated that "the evidence is in support of a general 'tragedy of the commons' when resources are held as open-access" (Berkes 1996: 89, 94), while another argued "open-access resources – those characterized by no property rights – will be overused, will generate conflict, and may be destroyed" (Ostrom and Schlager 1996: 128). As these statements indicate, "open access" replaces the "commons" as the problem. The very same scholars who argue for the importance of culture, institutions, and power in the commons simultaneously treat open access as lacking these factors that might shape how resources are used and allocated. Ignoring complexity *within* open access regimes, these common property theorists argue that economic rationality governs behavior in open access situations, such that individuals make purely economic decisions based on relative costs and benefits. In this view, then, property rights are essential.

Convergence

What all these seemingly different perspectives on the commons share is that they link forms of property, economic rationality, and environmental outcomes. Once common property theorists replaced the "tragedy of the commons" with the "tragedy of open access," the differences between what seemed like quite opposed positions are no longer so great. Without property regimes that constrain individual behavior, people will overcapitalize and overuse resources because it is economically rational to do so: this is the underlying argument of Gordon's model of the fisherman's problem, Hardin's model of the commons more generally, and the revisionist model of the benefits of the commons combined with the tragedy of open access. The solution, all argue, is to specify property rights in such a way as to limit access, provide economic incentives for conservation, and encourage exit from overcapitalized industries. It is these shared themes that link all these perspectives to neoliberalism, with its focus on market-based resource regulation through privatization.

Rationalizing the North Pacific

This convergence of thinking on property and the commons has influenced decision-making regarding fisheries management, as can be seen in the rise of neoliberal governance in the North Pacific Ocean. The North Pacific fishery, which includes the Gulf of Alaska, Aleutian Islands, and eastern Bering Sea, is by far the largest in the US, accounting in 2004 for 2.4 million metric tons, or one-third the national fish catch, and worth $1.2 billion. For well over a decade, fisheries managers, citing economic inefficiencies and "irrational" incentives of open access, have been developing plans for "rationalizing" all individual fisheries of this region, including those for halibut, pollock, king and snow crab, and possibly salmon (general information can be found at NPFMC 2006; statistics from NMFS 2005). Rationalization entails creating markets to govern resource use by enclosing fisheries within increasingly more delimited regimes of property rights, thus giving what was once a public good to a select group of fishers.

The North Pacific illustrates several points. First, the fact that fishery regulators have turned to specifying property rights as the primary means to achieve myriad management goals for this region is testimony to the pervasiveness of arguments about problems with open access and the promise of private property. Second, recent moves toward neoliberal fisheries management are not just spillover from neoliberal approaches more generally, but instead are situated within the longer history of emphasis on property and the problem of the commons. Third, neoliberalism can take a variety of forms when actually put into practice. The reality of these fisheries, combined with the history of fisheries management, leads to distinctive forms of neoliberal practice, including collective as well as individual forms of privatization.

For a decade after the US extended jurisdiction in 1976, fishery management of the North Pacific centered on developing, rather than limiting, US domestic fishing and processing capacity. By the late 1980s, however, regulators decided that there was too much capacity. To reduce capacity, they shifted toward rationalization through privatization in the form of limited entry, including a vessel moratorium followed by a license limitation program. These moves were seen as "an interim step toward a more comprehensive solution to the conservation, management, and economic problems in an open access fishery" (NPFMC 2002: 4). The first of these "more comprehensive" forms of rationalization was an Individual Transferable Quota (ITQ) program for halibut and sablefish, started in 1995. Halibut, especially, has long been an important fishery in the Alaska region, with catch in 2004 at 36,000 metric tons, worth $177 million. The most prominent form of privatization in fisheries, ITQs divide allowable fish catch among individual fishers, who can then either catch their quota allocation or sell it to other fishers. The goal of assigning these property rights is precisely to create a market – for access – where one did not exist before. ITQs reduce capacity by encouraging the least efficient vessels to voluntarily exit the fishery, for which they receive compensation by leasing or selling their quota. While building from this basic model, the halibut-sablefish ITQ program included numerous variations and protections. For example, it includes categories of quota for different kinds of fishers, and then placed limits on transfer of quota among these categories; there are also limits on the total amount of quota that any entity can control.

Following implementation of the ITQ, in 1998, federal regulators enclosed the fishery for Alaska pollock in a form of collective privatization. This fishery is the largest in the US with a 2004 catch of 1.5 million metric tons, worth $271 million. Privatization of pollock involved closing the fishery to all new entrants (qualifying vessels were named individually) and then dividing the annual total catch not among individuals, but among cooperatives of fishers and processors. Co-op members then decide how to further divide their allocation among individual vessels. Although not an ITQ, this type of enclosure works to reduce capacity in similar ways, as allocation can be shifted both within and between co-ops. Even though the co-op system uses collective ownership and decision-making, the underlying logic of the co-op system is the same as for ITQs: property rights are a market mechanism that can harness economic rationality to the goal of economic efficiency and conservation. Rather than devolving ownership and decision-making control to individuals or individual firms, this control is devolved to groups of firms. Thus, co-ops are consistent with neoliberal approaches to regulation, particularly in that they involve property-like mechanisms that create markets and switch governance from public to private control.

The crab rationalization plan, approved in 2005, incorporates aspects of both the halibut ITQs and pollock co-ops. Catch in the crab fisheries, which include king and snow crabs, in 2004 amounted to 21,000 metric tons, worth $142 million. As in the halibut-sablefish ITQ, initial allocation of the fishing quota would be made to individuals, with quota grouped into different regional categories. Also similar to the ITQ, there would be limits placed on consolidation and vertical integration. The similarity with pollock is that fishers could also form co-ops, in which they could easily shift their quota based on various formal and informal agreements among co-op members. The crab plan also adds a controversial new dimension to privatization, in that it includes transferable quotas for processors, such that those firms that are currently in the crab processing business would receive shares based on their processing history. While processors argue that this quota is necessary to protect their sunk costs in risky businesses in isolated areas, fishers fear that processing quotas will force them to sell all their fish to one or two customers, such that processors could dictate everything from timing of fishing to prices for the catch. In other words, many fear that private processor quota is a means of eliminating independent fishers, and replacing them with the equivalent of sharecroppers or piece-workers.

In addition to these existing programs, rationalization plans are also currently under development for the groundfish fisheries (cod, flatfish, rockfish) of the Gulf of Alaska, using the variety of mechanisms described for other fisheries, and there is a small-scale experiment in co-operative privatization of the lucrative salmon fishery of Alaska, which remains quite economically decentralized. These current efforts combined with the recent history of all fisheries of this region indicate that it is likely that the large-scale experiment in privatizing the billion-dollar-a-year fishing industry of the North Pacific has become entrenched.

Conclusion

These examples of the property revolution in one regional fishery in the United States show that property rights can take many forms, but that all of them revolve around this concern with economic rationality and enclosing the commons. These examples also show that enclosures of the commons can take the form of either individual or collective privatization. That neo-liberalism can encompass these varieties of privatization is also made clear by academic proponents of these plans, who do not make distinctions on individual/collective lines. Proponents of the benefits of the commons have recently argued for these privatization plans as long as they include some kind of collective decision-making – even if that collective is a group of firms (Holland and Ginter 2001; McCay 2001). In this view, using privatization to create market incentives can be consistent with the commons, as long as groups, rather than just individuals, are assigned rights. Several economists also suggest that it is unimportant whether property rights are collective or individual; rather, what is important is that privatization continue (Christy 1996; Pearse 1992). From all these perspectives, individuals acting collectively are much like any other single collective entity (e.g. a firm), and the use of property rights to enclose the oceans can proceed.

As these arguments show, common property theorists have been able to influence debates in fisheries by shifting the focus from the commons to problems more specifically with open access, and many economists and managers seem to accept the idea that some forms of common property management may be workable. But economists have been able to absorb this lesson without substantially altering their underlying argument about the relationship between property and economic rationality, problems with state-led regulation, or the importance of markets for environmental governance. To the extent that common property theorists focus on open access as inherently a problem, they are aligning themselves with orthodox economists' arguments about property, economic rationality, and state vs. market governance. As long as these theorists treat open access as a realm in which economic rationality prevails, rather than itself as a social relation in which different sorts of institutions and power relations are at work, they are limiting their critique of orthodox economic approaches; they more carefully specify existing models of social behavior and resource management, but do not offer completely different models that question assumptions of economic rationality and market behavior. The result is that even though these different groups of scholars seem to have quite different perspectives, they can all agree on plans for neoliberal privatization of fisheries to solve the economic and environmental problems that are assumed to result from open access.

It is in this sense that putting property at the center of fisheries problems is a neoliberal, market-based approach to ocean governance. All the approaches discussed in this chapter – whether private-, state-, or group-oriented – start from a particular economic logic that takes economic rationality as given. From this starting point, problems in fisheries are caused by open access regimes, which are a market distortion that inherently creates incentives to overuse, use inefficiently, race for the resource, and so on. Neoliberal privatization is seen to offer the solution, in that it creates market incentives that decrease capacity, increase efficiency, and encourage conservation (because each individual or group knows they can profit from the fish as much tomorrow as today). Market incentives may also lead to overfishing when "mining" fish stocks makes economic sense, and they will also cause "a high degree of real pain" among those who are not the beneficiaries of privatization, but this, proponents argue, is inevitable and inexorable, and all in the name of the greater good (Christy 1996: 288, 297). Property rights are at the center of a massive change in the political economy of the oceans around neoliberal, market-based socio-environmental policies that enclose for a few what was once the property of all. Neoliberalism in the oceans takes a particular form and has its own history and timeline based on the ways that, for over 50 years, fisheries analysts have structured regulation debates around the question of the commons and rationalization of the oceans.

Acknowledgments

A longer version of this chapter was Mansfield, B. (2004) "Neoliberalism in the Oceans: 'Rationalization,' Property Rights, and the Commons Question," *Geoforum* 35(3): 313–26.

References

Berkes, F. (1996) "Social Systems, Ecological Systems, and Property Rights," in S. S. Hanna, C. Folke and K.-G. Mäler (eds.) *Rights to Nature: Ecological, Economic, Cultural, and Political Principles of Institutions for the Environment*, Washington, DC: Island Press, pp. 87–107.

Berkes, F., Feeny, D., McCay, B. J. and Acheson, J. M. (1989) "The Benefits of the Commons," *Nature* 340: 91–3.

Caddy, J. and Cochrane, K. (2001) "A Review of Fisheries Management Past and Present and Some Future Perspectives for the Third Millennium," *Ocean and Coastal Management* 44: 653–82.

Christy, F. T. (1996) "The Death Rattle of Open Access and the Advent of Property Rights Regimes in Fisheries," *Marine Resource Economics* 11: 287–304.

Christy, F. T. and Scott, A. (1965) *The Common Wealth in Ocean Fisheries: Some Problems of Growth and Economic Allocation*, Baltimore, MD: Johns Hopkins University Press.

Crutchfield, S. R. (1986) "Extended Fisheries Jurisdiction in the USA: An Economic Appraisal," *Marine Policy* 10(4), 271–8.

Gordon, H. S. (1954) "The Economic Theory of a Common-Property Resource: The Fishery," *The Journal of Political Economy* 62(2): 124–42.

Hardin, G. (1968) "The Tragedy of the Commons," *Science* 162: 1243–8.

Holland, D. S. and Ginter, J. J. C. (2001) "Common Property Institutions in the Alaskan Groundfish Fisheries," *Marine Policy* 25: 33–42.

Johnson, C. (2004) "Uncommon Ground: The 'Poverty of History' in Common Property Discourse," *Development and Change* 35(3): 407–33.

Juda, L. (1991) "World Marine Fish Catch in the Age of Exclusive Economic Zones and Exclusive Fishery Zones," *Ocean Development and International Law* 22: 1–32.

Mansfield, B. (2001a) "Property Regime or Development Policy? Explaining Growth in the US Pacific Groundfish Fishery," *Professional Geographer* 53(3): 384–97.

—— (2001b) "Thinking Through Scale: The Role of State Governance in Globalizing North Pacific Fisheries," *Environment and Planning A* 33: 1807–27, with Erratum 34(1).

McCay, B. J. (2001) "Community-Based and Cooperative Fisheries: Solutions to Fishermen's Problems," in J. Burger, E. Ostrom, R. B. Norgaard, D. Policansky and B. D. Goldstein (eds.) *Protecting the Commons: A Framework for Resource Management in the Americas*, Washington, DC: Island Press, pp. 175–94.

Nadelson, R. (1992) "The Exclusive Economic Zone: State Claims and the LOS Convention." *Marine Policy* 16(3): 463–87.

Neher, P. A., Arnason, R. and Mollett, N. (1989) *Rights Based Fishing*, Dordrecht: Kluwer Academic Publishers.

NMFS (2005) *Fisheries of the United States 2004*, Silver Spring, MD: National Marine Fisheries Service.

NPFMC (2002) "Bering Sea Crab Rationalization Program Alternatives" (Initial Review Draft), Anchorage: North Pacific Fishery Management Council.

—— (2006) "North Pacific Fishery Management Council: Home. NPFMC", available at: www.fakr.noaa.gov/npfmc/default.htm.

Ostrom, E., Burger, J., Field, C. B. Norgaard, R. B. and Policansky, D. (1999) "Revisiting the Commons: Local Lessons, Global Challenges," *Science* 284: 278–82.

Ostrom, E. and Schlager, E. (1996) "The Formation of Property Rights," in S. S. Hanna, C. Folke and K.-G. Mäler (eds.) *Rights to Nature: Ecological, Economic, Cultural, and Political Principles of Institutions for the Environment*, Washington, DC: Island Press, pp. 127–56.

Pearse, P. H. (1992) "From Open Access to Private Property: Recent Innovations in Fishing Rights as Instruments of Fisheries Policy," *Ocean Development and International Law* 23: 71–83.

Scheiber, H. N. and Carr, C. J. (1998) "From Extended Jurisdiction to Privatization: International Law, Biology, and Economics in the Marine Fisheries Debates, 1937–76," *Berkeley Journal of International Law* 16(1): 10–54.

St. Martin, K. (2005) "Mapping Economic Diversity in the First World: The Case of Fisheries," *Environment and Planning A* 37: 959–79.

Steinberg, P. E. (2001) *The Social Construction of the Ocean*, Cambridge: Cambridge University Press.

5 Acts of enclosure

Claim staking and land conversion in Guyana's gold fields

Gavin Bridge

Mineral resources – gold, diamonds, or oil – are classically described as 'gifts of nature', the outcome of geological and hydrothermal processes that operate over time scales and at temperatures and pressures not replicable by society. These strong biophysical tethers make mineral resources analytically distinctive: like fish, water, game and land described in other chapters in this part, minerals exemplify Polanyi's (1944) category of 'fictitious commodities'. Although such distinctions are valuable, analyses which emphasize the physical provenance of ores and minerals can quickly end up naturalizing their status as socially-valued resources. This obscures the extensive socio-political work that must be done to commodify the invisible space of the underground and produce it as a bankable mineral deposit. Gold, oil and diamonds may be 'non-produced goods' from the perspective of classical political economy, but transforming the underground into an extractable, exportable commodity rests on the creation – and reproduction over time – of quite particular social relations. The 'resource space' must be nurtured as a site into and through which capital can flow, via knowledge claims which establish (and make legible) the mineralogical content of the underground, and by instituting property relations to the underground that enable its enclosure and the appropriation of its mineralogical values.

This chapter makes three arguments concerning the neoliberalization of access regimes to mineral resources and illustrates these arguments by reference to Guyana. First, the widespread adoption of neoliberal investment codes for mineral resources extends a long history of enclosing 'public natures' and their private appropriation as nature-based means of production. The sovereign mineral estate historically has been among the first of the various public realms of nature (lands, forests, waters, species, environmental services) to which private rights are assigned: appropriating the mineral kingdom exemplifies the process labelled by Marx as primitive accumulation.[1] The actions of sovereign powers to assert dominion over the 'vertical territory' of the underground, and subsequently to enclose it for private gain, are closely associated with the birth of modern capitalism (Braun 1997; Mumford 1934). Enclosing the underground, then, is a long-established process and not one that is unique to neoliberalism.

Second, the distinctiveness of neoliberal modes of governing the subterranean mineral realm lies in the way they seek to transform the underground into a *site for the circulation of capital* (and, in particular, for the circulation of international capital). In most cases neoliberal mineral policy reforms have not introduced the 'resource imaginary' or ushered in mineral extraction: many of the African, Latin American and Asian economies that liberalized in the 1990s had active (frequently state-owned) mining industries in the pre-liberalization period. Rather, the key transition involves turning the underground into a site for the circulation of capital raised in metropolitan centres like Toronto, Johannesburg or London. Neoliberalism, therefore, differs from other modes of resource governance – such as those based on concepts of commonwealth and/or national patrimony, or those which articulate natural resources as a foundation for import-substitution industrialization – in its attempt to turn the highly localized underground 'inside out' by attaching it to the strategic objectives of non-localized investment capital. This instrumental coupling of radically different geographical scales is a distinctive feature of neoliberal mineral policy reforms and is expressed in the institutional form of enclosure.

Third, the form and scale of these institutions have material consequences on the ground. As advocates of an 'internationally competitive' mining regime for Guyana in the early 1990s were well aware, the institutions through which mining laws are implemented strongly influence the rate and extent of mining investment in Guyana. By determining when, where, how, and with what compensatory provisions (if any) mining may take place, institutions of enclosure condition the ways in which policies of economic liberalization and mineral promotion impact people and environments.

Each of these arguments hinges on the *mine claim*, an institution of property through which individuals and firms gain access to the underground. The chapter takes the claim as its point of departure, and proceeds analytically first *upstream* from the claim to understand the production of the claim as an institutional re-scaling of the Guyana subsurface with the intent of attracting foreign investment; and then *downstream* from the claim to illustrate how claim provisions came to produce exploitable resource spaces in the hinterland of Guyana, driving land use change and land conversion (Figure 5.1).

Liberalization: opening the underground

In the ten year period beginning in 1985, over ninety states adopted new mining laws or revised existing legal codes in an effort to promote foreign investment in their mining sector. The promulgation of new Mining Codes was frequently part of a broader package of neoliberal administrative and fiscal reforms. Their combined effect was to open up new opportunities for the international mining industry in areas that were formerly either closed *de jure* because of political restrictions, or closed *de facto* since political-economic risk was sufficiently high to deter prudent investment.

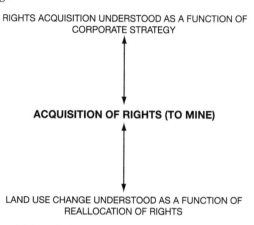

Figure 5.1 The acquisition of rights as a means of coupling economic processes and environmental outcomes.

Inside out: promoting Guyana's natural resources

Like many other developing countries, Guyana pursued autarkic economic development policies during the 1970s and 1980s. These favoured extensive state intervention in the economy – including nationalization of sugar, bauxite and timber assets – and restrictions on the role of private, and especially foreign, investment: by 1976 state control of the economy exceeded 75 per cent. In the late 1980s, however, declining raw material prices and mounting external debts propelled a series of neoliberal economic reforms as part of an IMF-supervised programme of structural adjustment. The country adopted an Economic Recovery Plan (ERP) in 1988, incorporating economic stabilization, liberalization of investment and foreign exchange regimes, privatization of former state holdings, and divestiture of state lands (Lemel 2001). Stimulation of natural resource exports and the attraction of foreign investment in natural resource production were central elements of the ERP. Natural resource exports (sugar, rice, gold and timber) increased following adoption of the ERP, and gold mining played an important role in the country's economic revival during the mid-1990s, a period in which GDP rose an average of 6 per cent per year (NDS 1996). Foreign ownership of businesses was authorized in 1988 and a number of significant investments were made in timber and mining during the 1990s. By the end of the decade, sugar, rice and gold were the principal foreign exchange earners for the country, with around 40 per cent of Guyana's merchandise exports being comprised of rice and sugar, 26 per cent of gold, and 8 per cent of unprocessed forest products.

A gold mining boom

Of particular significance to the gold and diamond mining industry were the passage of a new Mining Act in 1989 (which improved access to exploration and mining rights), and the liberalization of the role of the state gold-buying agency – the Guyana Gold Board – which, up until the early 1990s, had operated as part of a strict exchange control regime. Such legal and fiscal reforms were an explicit attempt to promote mineral development and facilitate foreign investment in the sector by changing the perceived risk/reward ratios for investing in Guyana's natural resources. The new Mining Act created incentives for large and small-scale producers to invest in mining and coincided with a period of relatively buoyant international gold prices. Reported gold production increased 10-fold during the 1990s, and by the end of the decade gold was accounting for approximately 20 per cent of GDP and over 25 per cent of exports.[2] The reforms of the 1990s restored a level of gold production in Guyana that had not been seen since the 1890s, when (then) British Guiana was among the top 10 most prolific gold-producing regions in the world (Swain 1980): prior to the mid-1990s – when reported gold output reached 13,500 kg per year – the peak year for gold production in Guyana had been 1893 (4,300 kg).

Over half of the increase in gold output since 1990 may be attributed to the commissioning of a single large-scale, hardrock mine at Omai on the Essequibo River. Attracted by both the country's geological potential and its newly-competitive mineral investment regime, the Canada-based firms Cambior and Golden Star opened the US$300 million hard-rock mining operation in 1993. Until its closure in September 2005 Omai ranked among the largest gold mines in South America, accounting for around 70 per cent of Guyana's formal gold output. Omai, however, has been an exception within the Guyanese gold mining industry: for much of the twentieth century, gold mining in Guyana has consisted principally of small and medium-scale producers working placer deposits. These vary in their level of capitalization and the technologies employed but are typically either hydraulic operations extracting gold from alluvial floodplain deposits or suction dredge operations working the riverbeds: gold recovery in both cases is typically through a combination of gravity concentration and mercury amalgamation. These small and medium-scale producers contribute significantly to the country's gold output: reported production by small and medium-scale producers rose from 539 kg in 1989 to 3,425 kg in 1996. The 1990s saw an increase in the number of dredge licences, as well as a very large expansion in the geographical area being accessed for mining under mine claims, permits and exploration licences.

Institutions of enclosure

National efforts to attract mineral investment from metropolitan centres like Toronto, London or Johannesburg have centred on changing the terms under which individuals and firms can gain access to the mineral estate. Drafters of the Guyanese National Development Strategy were acutely aware of the need to compete with other mineral-rich countries seeking to secure inward investment in their mineral properties:

From west to east, there is a global preoccupation with advertising one's mineral heritage, revising mining laws and fiscal policy, in some instances, offering fiscal incentives; advertising one's mining culture, infrastructure, educated human resources, etc., as additional attractions to the potential investor. In some ways it is like jostling for attention in a crowded marketplace.

(National Development Strategy 1996)

In practice, 'jostling for attention' has focused on reworking the institutions that allocate private rights for mineral exploration (exploration licences) and for mineral extraction (mine claims).[3] Exploration licences are the primary means by which firms establish the mineralogical prospectivity of the underground and identify the most attractive targets for investment: exploration licences therefore convey an exclusive right to prospect and to 'cherry-pick' the best locations. Mining claims are the means by which individuals or firms gain access to – and demonstrate secure tenure of – gold-bearing deposits. Both exploration licences and mine claims are institutions of property, bundles of rights that convey to the holder rights of access to a specific parcel of land for the purposes of either exploring for or extracting minerals. In most jurisdictions exploration licences and mining claims do not convey title to the material in the sub-surface, but provide a time-limited monopoly right to explore for or extract minerals. Guyana's Mining Act (1989) affirmed sub-surface mineral rights were vested in the state, but liberalized the terms and conditions under which the Guyana Geology and Mines Commission could grant access to the mineral estate to private entities. Exploration licences and mining claims, then, can be thought of as 'technologies of state power' that rationalize the mineral estate through a combination of measurement, calculation, recording and visualization, and which collectively perform the resource as an exploitable space (Peluso and Vandergeest 2001; Mitchell 2002). Licences, claims and the mining cadastre are also narrated as appropriately *neoliberal* technologies of state power: they establish the underground as a terrain of political regulation by affirming property rights and providing information to the market (on, for example, the location of claims).

Guyana became a prime target for exploration in the early 1990s. Relatively buoyant gold prices created a bullish environment for exploration, with firms seeking to acquire rights to potential 'hot spots' in places like the comparatively under-explored Guiana Shield. The successful development of the large-scale Omai mine in 1993 led other firms to seek out 'Omai-type' targets in a region widely regarded as hosting several multi-million ounce gold deposits. Major discoveries in Ghana on the geologically-analogous West African Shield further indicated the possibilities for significant finds within Guyana. 'Junior' and 'senior' mining companies competed with one another to identify and lock up the rights to 'world class' deposits.[4] For senior mining firms (such as Placer Dome, BHP, Golden Star and Cambior) such deposits – with their relatively low operating costs, long-life reserves, and ability to generate substantial profits over the life of the mine – would help secure their competitive position by significantly increasing their reserves and providing the basis for long-term profitability. For 'junior' mining firms (such as Vannessa, Romanex, and Guyana Goldfields) that entered Guyana during the 1990s, the discovery of a multi-million ounce gold deposit would dramatically drive up their share price, increase their access to capital, and enable them to make a substantial profit by either selling the property or joint-venturing it with a more established, senior firm.

Thus, from 1990 to 1996, Guyana played host to both senior and junior mining firms, some of which acquired exclusive reconnaissance rights to large areas of the country in an effort to locate significant resource targets. The fall in gold prices after 1996, however, drove companies to cut their exploration budgets and adopt more cautious positions regarding the risks/ rewards of potential investments.[5] As a result, exploration activity in Guyana was preferentially abandoned in favour of more established regions such as Canada and Australia, as well as other South American countries (like Chile and Peru) that were perceived as more prospective and/or less risky locations. The waxing and waning of exploration activity within Guyana during the 1990s illustrates how geographies and histories of investment in a particular place are connected to non-place-specific trends, such as the volatility of commodity prices, the availability of project finance, or inter-firm competitive strategies.

Scaling access to the underground

While the rate, pattern, and manner of mine claims acquisition can be related to corporate strategy, the claim itself – i.e. rules and norms describing the way land may be accessed for mineral exploration and extraction – can be understood as a codification of prevailing social relations, the result of political struggles and compromises between different interest groups over time. In Guyana, the institutional form of exploration licences and mining claims illustrates a tension between international lending organizations promoting access for foreign investment (for example, via calls for non-discriminatory treatment of domestic and foreign investors) and domestic interests concerned to keep mining – and more specifically, rights to land – under Guyanese control. These struggles have resulted in a distinctive mining regime: while it shares commonalities with many other liberalized investment regimes around the world, the configuration of Guyana's mining institutions reflects the compromises made between different political and economic interests. In this sense, national mining legislation can be understood as a particular scalar codification – a scalar fix – of a normative struggle over how the economic space of Guyana (including heterogeneous spaces *within* Guyana, such as the hinterland vs. coastal zone or urban vs. rural areas) should articulate with the 'global economy'. It is possible, therefore, to read specific features of the law – such as differentiation/ discrimination between domestic economic interests (who can own mine claims and permits) and foreign economic interests (who cannot) – as efforts *to scale* economic processes so as to achieve particular outcomes. For example, the introduction of the medium-scale permit provision in 1993 (see below) was a conscious attempt to circumvent limited sources of investment within Guyana by scaling up investment flows, both in the geographical sense of 'going global' for sources of finance as well as the more conventional sense of increasing the volume/value of investment flows.

Liberalization and land conversion

The extractive nature of mining means that the right to mine is also a right to convert land from one use to another and to use water for mining purposes. To the extent that it allows the miner to discharge mined materials into the environment, mine claims also confer a right to pollute. Conditions attached to mine claims have the potential to determine *when* (e.g. seasonal prohibitions?, regulation of the intensity of mining over time?), *where* (e.g. on Amerindian lands?, mining of river beds and banks?, near reserved lands?), and *how* (e.g. specification of mining practices?, requirements for managing mercury use?, backfilling of pits?) mining may take place. Thus the form and scale of the institutions of enclosure strongly influence the ways in which economic liberalization and efforts to attract mining investment will impact existing land uses. In short, as an institution mediating the way in which investment drives local environmental change, institutions of enclosure like mine claims are not only *differentiation made* – an historically and geographically specific product of social processes – but also *make a difference* in that they actively shape the environmental outcomes of investment.

Figure 5.2 substantiates this argument by demonstrating the material effects of mining law reforms in Guyana in terms of the areal extent and geographical distribution of mine claim activity in the six mining districts in Guyana for the period 1984–2001. The liberalization of mining laws in 1989 helped drive an increase in new claim activity: the number of new claims more than doubled from 1,316 in 1988 to 3,070 in 1991. This was a period in which gold prices actually *fell*, suggesting that the increase in claim activity in Guyana was driven more by the liberalization of mining legislation (which facilitated the process of making claims) than by international price trends. While commodity prices clearly have some bearing on investment activity, the volume and geographical pattern of investment flows are strongly conditioned (facilitated or retarded) by the institutional environment within which investment is embedded.

Figure 5.2 demonstrates how the land area permitted or claimed for mining each year rapidly expanded after 1992 when a new regulatory provision was introduced authorizing the Geology and Mines Commission to issue mining permits for areas of up to 1,200 acres. This created a new

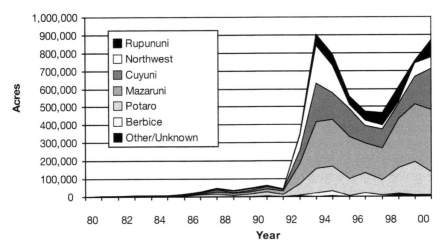

Figure 5.2 Land area of new mine claim and mine permit applications by mining
district, 1980–2000.

class of mineral land holding – the 'medium-scale permit' – to supplement
the long-standing, 21-acre 'small-scale claim', the basic mineral property
right in Guyana. The medium-scale provision was central to efforts to create
an internationally competitive investment environment capable of attracting
inflows of capital to the gold and diamond mining industry. With domestic
investment limited in availability, the government of Guyana sought to
attract investment from non-Guyanese entities as a way of increasing gold
production and upgrading exploration and production technology. Although
the new permits – like claims – could only be owned by Guyanese, their
longer tenure period (5 years vs. 1 year) and their considerably larger size
(1,200 acres vs. 21 acres) made them more attractive than claims as a means
of leveraging investment via joint-ventures between Guyanese mine permit
holders and external sources of capital.

Introduction of the provision, therefore, indicates how liberalization involved
explicitly scalar strategies on the part of the state: by re-scaling the bundle of
property rights (i.e. conferring these rights to larger areas of land for longer
periods of time) that are fundamental to mining activity, the state sought to re-
scale capital flows into mining in Guyana by enabling domestic permit holders
to access transnational sources of capital, technology, and expertise. Reform
efforts sought, in other words, to reposition Guyana as an annex to the global
economy, or more specifically as an outlet for investment capital raised on the
principal mining exchanges in Toronto and, to a lesser extent, London.

Figure 5.2 also suggests how this re-scaling of economic processes and
institutions can have material effects. The introduction of the medium-scale
provision in 1993 drove a twenty-fold expansion in the area of land claimed
or permitted for mining as individuals sought to acquire the rights to mine.
On the ground, this expansion occurred as a process of enclosure in which

private property rights (for mining) were assigned to lands formerly vested in the state. The net effect of the annual permit activity illustrated in Figure 5.2 is that an extensive area of land has come under private control for mining during the 1990s: one source estimates that the area under claims/permits increased from 200,000 acres to around 3 million acres between 1990 and 1994 (Forest Peoples Programme 2000: 47). Figure 5.3 illustrates the variation in intensity of mine claims across the country and the way in which mineral promotion has created a geographically uneven landscape of investment and enclosure. Each square on the map represents a 1:50,000 topographical sheet (which covers an area of approximately 231,000 acres), and the map illustrates how up to 100 per cent of the land area has been claimed for gold and diamond mining in parts of the Mazaruni and Cuyuni Mining Districts. Analysing and mapping histories and geographies of property institutions during liberalization, then, provides an empirical context through which to connect economic geography's conventional interests in processes – demonstrating, for example, how liberalization is achieved through the re-scaling of economic processes and institutions – with a demand for analyses capable of demonstrating the significance of these processes in terms of the outcomes they generate on the ground.

To take this one step further, what have been the effects on land use of the rapid expansion in the area claimed for mining? While gold production increased during the 1990s in Guyana – an increase that has been accompanied by concerns about its effect on water quality (increased sediment loads and the potential for mercury contamination) and its role in facilitating the spread of malaria – growth in mine output lags considerably the increase in permit activity.[6] This suggests that much of the land enclosed by mine permits is not actually being mined and, therefore, that a substantial amount of permitting activity is speculative. Discussions with knowledgeable sources in Guyana support this assertion – much of the expansion of claim/permit activity has *not* been accompanied by active mining – and point to an emergent problem of landlordism as a relatively small number of individuals have acquired the rights to significant areas of mineralized land that is then not mined. This lends support to the argument that the rapid expansion in the area claimed or permitted for mining has actually *slowed* potential mining activity by creating incentives to acquire – and retain – land which is then effectively unavailable for others to mine.

Explanations for this somewhat contradictory outcome can be linked back to the particular ways that processes of liberalization played out in Guyana. Liberalization is a complex scalar process that did not simply 'open up' Guyana as a global capitalist space by eroding the authority of the state. The 'global capital flows' that linked the Toronto Stock Exchange to the auriferous and diamondiferous gravels of the Upper Mazaruni or Potaro were dependent on action by the state to clarify and enforce (via the extension of property rights) the ability of individuals to control the micro-territorial spaces of the mining claim. Thus liberalization simultaneously re-affirmed Guyana as a national space via, for example, stipulating that Guyanese citizenship be required in order to own mine claims while also seeking to transnationalize it. And it is the intersection of these two scalar processes that drove the largely speculative acquisition of mining rights: by enabling entry of foreign investment but denying the possibility of foreign ownership of permits and claims, the mining law created an opportunity for

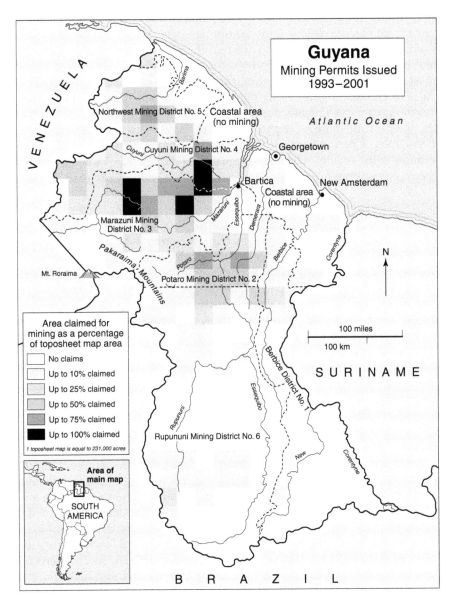

Figure 5.3 Location and intensity of mine permitting in Guyana, 1993–2001.

local actors to derive rent from the exclusive property rights afforded through a mine permit, by entering into joint ventures with external sources of capital. It is important not to overstate the case here: many holders of claims and permits seek to realize the economic value of these rights by working permitted areas for gold and/or diamonds. Some evidence for this can be found from a review of the number of applications for licences to operate mining dredges: new applications increased by over 75 per cent from 168 in 1992 to 285 in 1994. The point, however, is that the potential rents from entering into a joint venture are considerably greater than that from mining, creating an incentive (given the relatively low cost of acquiring and holding claims and permits) to acquire the rights to large areas of the most promising land with the objective of leveraging, via a joint venture, the value of exclusive rights mining rights at a future date. Records of firms that began operating in Guyana during the early 1990s, for example, suggest that the holders of mineral rights could be paid between US$3,000 and US$4,000 per permit for their role in putting up the land as part of a joint venture, with royalty agreements capable of earning the holders tens of thousands or even hundreds of thousands of U.S. dollars per year.

Gifts of nature as sites for circulation

Guyana exemplifies many processes that are common to developing countries: international pressures (via structural adjustment) to adopt policies of economic liberalization such as export promotion, the attraction of foreign investment, reductions in state spending, and privatization of key sectors; incomplete and sometimes contradictory domestic legislative efforts to pursue these economic reform objectives; and sporadic inflows of foreign capital into leading export sectors.

A neoliberal-inspired process of mineral promotion in Guyana has enabled some individuals and firms to connect land parcels to external sources of investment. This has increased the effective value of these lands for mining, and has also driven (unintended) speculative activity by others who also seek to realize an increase in the value of land by connecting nationally-endowed mining rights with sources of capital from outside Guyana. Although much of the value of land is speculative – in the sense that the value of these lands is currently not being realized and is contingent on conditions being met in the future – it is the prospect of changes in land values that has driven the rate and pattern of claims/permit acquisition and which, by extension, drives changes in land use. This valorization (or de-valorization) of land leads to a shuffling in the socio-spatial distribution of rights to use land – as illustrated in Figures 5.2 and 5.3 – which have the effect of embedding individual land parcels within different sets of political-economic incentives. For example, a mining claim/permit acquired from the government by a small-scale operator with relatively few other sources of income may be actively mined to generate an income stream (driving a set of environmental impacts related to clearing tropical forest cover and working the land for mineral extraction) while other claims/permits may be acquired and held (and not actively worked) as a speculative venture by individuals interested in maximizing the potential future value of claims.

Zimmerman's phrase 'resources are not, they become' is an enduring aphorism within resource geography (Zimmerman 1933). His perceptive claim that natural resources are social constructs has subsequently been much sharpened by insights into the production of 'capitalist nature', and the recognition that producing minerals *as resources* is a fundamentally political process: materials must be 'coaxed and coerced' from landscapes where they already may be valued in quite different ways (Tsing 2004; Bakker and Bridge 2006). This chapter has sought to illustrate a specific moment in the making of mineral resources in Guyana – the production of the underground as a site for the circulation of international capital during the 1990s. It has argued that it is not the act of enclosure that marks this moment as distinctive, but the way in which institutions of enclosure sought to render the subterranean spaces of Guyana legible and attractive to investment capital raised in metropolitan centres like London or Toronto. The elusive glint of gold in the river gravels of Guyana would seem to confirm that minerals are indeed 'gifts of nature': however, producing these minerals as targets for foreign investment and enlisting them in a neoliberal model of development required institutional interventions to re-scale the spaces of the underground.

Acknowledgements

A National Science Foundation Career Grant (SBR #9874837) supported research on the environmental implications of direct investment in the mineral sector. A grant from the National Geographic Society (Grant #6804–00) supported earlier field experience in Guyana. I thank the Guyana Geology and Mines Commission for providing access and staff time to assist the collection of data on mine claim and mine permit histories in Guyana. Gerardo Castillo provided research assistance and Geoff Maas drew the map. Acknowledging the above implies no responsibility for errors of omission, commission or interpretation which remain those of the author alone.

A longer version of this chapter was Bridge, G. (2002) 'Grounding Globalization: The Prospects and Perils of Linking Economic Processes of Globalization to Environmental Outcomes', *Economic Geography* 78(3): 361–86.

Notes

1 The conquistadors' seizure of land, people and resources in their predations on New Spain during the sixteenth and seventeenth centuries are described by Marx ([1867] 1990: 915) as being among the 'idyllic proceedings (that) are the chief moments of primitive accumulation'.

2 Reported production may significantly underestimate the role of domestic gold producers.

3 In Guyana the term 'mine claim' has a particular meaning and refers to the rights and responsibilities associated with a small-scale mining claim, as distinct from those related to a medium-scale permit or large-scale licence. This contrasts to use of the term in everyday speech where it refers to a generic category of mine-related land holdings.

4 The distinction between junior and senior firms in the non-ferrous metals mining sector refers primarily to their different degrees of market capitalization. It also represents a convenient shorthand for the different strategic objectives of these two classes of mining firms. Juniors are typically focused on exploration (chiefly for gold) and drive increases in shareholder value through the discovery and proving of precious metals deposits. They tend to have a more aggressive approach to risk/reward ratios than do senior firms and are better positioned to capture the value from smaller deposits. Senior firms also undertake exploration, but typically hold a larger range of assets by both commodity and geographic setting (they may also control downstream processing stages for base metals, like smelting and refining for copper and nickel) and tend to focus on adding value through advantages in technology, management and access to capital.

5 Gold prices remained relatively low (below US$300 per ounce) until mid-2003.

6 Rushes or 'shouts' in the Mazaruni and Potaro Districts have been associated with localized expansions of mining activity.

References

Bakker, K. and Bridge, G. (2006) 'Resource Regulation', in K. R. Cox, M. Low and J. Robinson (eds) *Handbook of Political Geography*, London: Sage.

Braun, B. (1997) 'Buried Epistemologies: The Politics of Nature in (Post)colonial British Columbia', *Annals of the Association of American Geographers* 87(1): 3–31.

Forest Peoples Programme (2000) *Indigenous Peoples, Land Rights, and Mining in the Upper Mazaruni: A Report by Upper Mazaruni Amerindian District Council*, Amerindian Peoples' Association of Guyana and the Forest Peoples Programme.

Lemel, H. (2001) *Patterns of Tenure Insecurity in Guyana*, Madison, WI: Land Tenure Center, Working Paper no. 43, University of Wisconsin-Madison.

Marx, K. ([1867] 1990) *Capital*, vol. 1, New York: Penguin.

Mitchell, T. (2002) *Rule of Experts: Egypt, Techno-Politics, and Modernity*, Berkeley, CA: University of California Press.

Mumford, L. (1934) *Technics and Civilization*, New York: Harcourt Brace.

National Development Strategy for Guyana (1996) Chapter 32 – Mining Policy. Online. Available at: www.guyana.org/NDS/chap32.htm (accessed 8 May 2006).

Peluso, N. L. and Vandergeest, P. (2001) 'Genealogies of the Political Forest and Customary Rights in Indonesia, Malaysia, and Thailand', *Journal of Asian Studies* 60(3): 761–812.

Polanyi, K. (1944) *The Great Transformation*, Boston, MA: Beacon Press.

Swain, W. (1980) 'The Gold and Diamond Mining Industry of Guyana', MSc degree thesis, Imperial College of Science and Technology, University of London.

Tsing, A. (2004) *Friction: An Ethnography of Global Connection*, Princeton, NJ: Princeton University Press.

Zimmerman, E. (1933) *World Resources and Industries*, New York: Harper and Brothers.

Part I

Commentary

6 Enclosure and privatization of neoliberal environments

Nancy Lee Peluso

The chapters in this part deal with the production of new sorts of fictitious commodities from a variety of "natures" newly privatized and circulated under neoliberal reforms. As a collection, they resoundingly demonstrate that enclosure is alive and well in the twenty-first century. The authors vary in the ways they refer to, conceptualize, and address enclosure – from "primitive accumulation" to "dispossession" to "privatization" to creating new "property rights" and "claims." This speaks to an unresolved tension vis-à-vis the relation between enclosure and primitive accumulation. Yet, all the chapters engage productively with a relatively new debate about the historicity or trans-historicity of dispossession, primitive accumulation and accumulation (e.g., de Angelis 2001; Zarembka 2002). All show that contemporary enclosures relate or react to historical forms of enclosure and claims-making in the sites or circumstances they take as their subjects.

One of the take-home messages from this part is that economic processes are hardly disembedded from the states, social relations, or environmental contexts in which they occur. This is a central problem with the orthodox neoliberal trope of "free" markets. In demonstrating the range of forms of state involvement, these chapters all challenge (and will hopefully help put to rest) what Swyngedouw calls "one of the central myths of the neoliberal model, i.e., that privatization means getting the state off the back of the economy and rolling back regulatory red tape" (pp. 55–56). In fact, a commonality of enclosures is that they inevitably involve some kind of public–private or state–capital alliance. For example, the concern to establish property rights in resources, associated with neoliberal approaches to environmental governance, is derived from the need to legitimate practical enclosures through laws, enforced by state agencies. This is true whether private, common, corporate, or state property rights are sought (Mansfield). Thus although state power may have changed in form, scale, type of practice, or effectiveness, it remains constitutive of neoliberal schemes and regimes.

This requires some nuance in how states are understood. States, as complex, shifting, and contingent assemblages of institutions, actors, policies and laws, are engaged in balancing acts between enabling access to resources and yet also limiting it. Gavin Bridge engages state power and form by unpacking the state institution he calls "the mining claim," created as part of nationa-lized ownership of all resources and extraction enterprises. Similarly, in cases of water provision and fisheries management, Swyngedouw and Mansfield, respectively, demonstrate that states retain key roles in governance and derive rents through technical institutions. These mechanisms enable state agencies and other in-country actors to engage in negotiations with corpo-rate and other private interests. In a different set of configurations, Robbins and Luginbuhl argue that "eco-managerial bureaucracies" such as the state wildlife bureaus in Montana, use what they call their "natural" alliances *with* local hunters and *against* private enclosures by in- and out-of-state elites. Meanwhile, NAFTA and other trade agreements actually call states to task for not enacting transparent laws designating which state authorities are responsible for management of specific resources (McCarthy). In each of these diverse cases, *re-regulation* is more evident than *de-regulation* simply understood. The chapters go beyond this, however, showing what is regu-lated anew, who benefits, and how various neoliberal reforms are legiti-mated by ideologies, truth claims, and discursive and institutional strategies.

Through their re-organization and re-orientation, the persistence of state institutions can be seen as Polanyian responses, i.e., "double movements" (2001: 75–6) that emerge out of the struggles between states and capital. Polanyi's double movements of capital and state practice underlie McCarthy's analysis of neoliberal capital's efforts to enclose the conditions of produc-tion, even in the context of making trade agreements. In Swyngedouw's recap of the history of urban water delivery, we see the dialectical relationship between state and capital from one historical "stage" of water provision to another. These two cases and that of Guyana– where state controls over mining claims are not eliminated but altered under the new regime – high-light the importance of "putting the state back in" (Evans *et al.* 1985) and understanding new or "graduated" forms of sovereignty (Ong 2000).

Yet, what remains puzzling in regard to state actions are the complex motives and mechanisms by which state actors agree to these arrangements and either purposely or by default renege on the protections of their citizens and territories that historically imparted legitimacy to their rule. In other words, we need to understand what has happened to the "social" in these new social relations of enclosure and what the "neoliberal" in "neoliberal environments" has to do with its sidelining or elimination. The overall pic-ture presented in these chapters is not one of state institutions protecting their citizens and territories, but rather of their establishing regulations to gain or maintain a piece of their sale. McCarthy points out that the new trade agreements castrate other, more protective, realms of international and national law, giving rise to the nagging question of "why?" Granted, IMF restructuring programs create severe conditions and terms that must be met, but some of these arrangements are so draconian as to threaten the very survival of their citizens and the environments within their territorial bounds. This case in particular challenges the neoliberal argument that private property rights are more conducive to sustainable management – largely because the intent of these agreements is to remove the social constraints on capital accumulation, including environmental and human rights protections. The agreements thus negate the claims of national citizens in the short term and threaten their own long-term survival

Besides excluding "the social," the unintended consequences of enclosure seem to manifest in onslaughts on the enclosed natures themselves. Enclosure's macabre effect in Robbins and Luginbuhl's chapter is the degradation – through Chronic Wasting Disease – of the nature whose value was meant to produce surplus, i.e., the bodies of the enclosed wildlife kept in "unnaturally" close proximity or shipped across state lines into territories they did not earlier inhabit. This kind of negative feedback has parallels in the higher incidences of habitat-poisoning that can result from shrimp aquaculture (Stonich and Vandergeest 2001), or the pollution of someone else's nest as the Metalclad company did in reopening a closed waste dump in Mexico (McCarthy). Environmental stresses play out in other ways in the water privatization, fisheries, and mining cases, where initial efforts to "rationalize" production through property rights all led to increases in production and not to more stringent protections.

The neoliberal-era enclosures discussed here are all related to longer histories of enclosure and capitalism. The maintenance of components of both private/liberal and statist/Keynesian water provisioning systems (Swyngedouw), or the allocation of private property rights to fisher collectives (Mansfield) hearken back to local and resource/commodity-specific histories. The tensions between McCarthy's and Bridge's chapters illustrate how histories of state–capital relations can differentially influence neoliberal international agreements. In the former chapter, Mexico's nationalist history and practices in the post-WWII developmentalist decades energized NAFTA framers to preclude nationalization of foreign enterprises in the new investment agreements. On the other hand, in Bridge's analysis of Guyanan mining claims, nationalism and a history of nationalizations drove the Guyanans to bargain hard around questions of citizenship and property rights.

How capital has responded to nationalist constraints is of course central to US history as well. The nineteenth-century Acts producing enclosures in the American West compare well with neoliberal-era enclosures because they were enacted during a global period of liberalism and *"laissez-faire* capitalism." The ones mentioned by Robbins and Luginbuhl were intended to encourage smallholder settlement of the arid and less productive (for intensive farming) western lands; state and settlers both benefited. Other Acts of that period facilitated capitalist production or state enclosure of forests (e.g., the Timber and Stone Act and the Organic Act). The Lode Law and General Mining Act gave mining interests access to public lands, arrangements that worked first in the interests of small-scale miners and later to the significant benefit of large, highly capitalized firms by providing almost rent-free access.

The authors' engagements with multiple scales of social relations embedded through history highlight the layered multiplicity of interests and sometimes surprising beneficiaries of new enclosures, and are a reminder that neoliberal reforms are politically constituted and contested. As a collection, these pieces provide a diverse picture of the variety of struggles and gains connected with new enclosures. Moreover, as the vast literature on common property resources has shown, management is often about the "social" dimensions of property and enclosure. Lifestyle choices, magical-religious connections to land, forests, fish, and the sea, status value, history and memory, and other social meanings and relations often drive both collective and individual resource ownership in ways that have no correlate in neoliberal policy – or analyses that ignore these social meanings. In this regard, the Montana case provides a glimmer of hope for maintaining a measure of the social in resource production, in part because of the social meanings of wildlife, and despite the histories of enclosures in the West. In some instances, therefore, capitalist, smallholder, and subsistence users' interests are all served by state policy and practice or by changing public–private alliances.

What do these apparent contradictions tell us more broadly about "neoliberal environments" and the privatization of nature? On the one hand, as Bridge notes, enclosures may be "differentiation made" but they also generate differential effects because of their very social and environmental embeddedness. We have seen that the effects can range from new sorts of property relations to new alliances and other practices to facilitate informal access, reminding us that the pronouncements of neoliberalism's advocates are largely unproven.

Further, the newness of most of these fictitious commodities, and their situated emergence out of neoliberal political economies, provoke important theoretical questions about the ongoing nature of enclosure and primitive accumulation (de Angelis 2001, 2004; Perelman 2000; cf., Zarembka 2002). They also suggest possibilities for challenging neoliberal policies, practices, and institutions, even without depending on civil society-cum-NGOs. State institutions remain a fulcrum point where changing policies can affect the distribution of access to resources and sustain environments and livelihoods. At the same time, scholars and critics of neoliberalism have yet to answer an intriguing puzzle: given the importance of history, politics and social embeddedness, how and why did state and civil society institutions allow such widespread neoliberal enclosures – representing so few people's interests – to proceed?

References

De Angelis, M. (2001) "Marx and Primitive Accumulation: The Continuous Character of Capital's 'Enclosures,'" *The Commoner* 2 (September), available at: www.thecommoner.org.uk/02deangelis.pdf.

—— (2004) "Separating the Doing and the Deed: Capital and the Continuous Character of Enclosures," *Historical Materialism* 12(2): 57–87.

Evans, P.B., Rueschemeyer, D. and Skocpol, T. (eds) (1985) *Bringing the State Back In,* Cambridge: Cambridge University Press.

Ong, A. (2000) "Graduated Sovereignty in Southeast Asia," *Theory, Culture & Society* 17(4): 55–75.

Perelman, M. (2000) *The Invention of Capitalism: Classical Political Economy and the Secret History of Primitive Accumulation,* Durham, NC: Duke University Press.

Polanyi, K. ([1944] 2001) *The Great Transformation: The Political and Social Origins of Our Times,* Boston, MA: Beacon Press.

Stonich, S. and Vandergeest. P. (2001) "Violence and the Blue Revolution: Industrial Shrimp Farming in Thailand and Honduras," in N. L. Peluso and M. Watts, *Violent Environments,* Ithaca, NY: Cornell University Press.

Zarembka, P. (2002) "Primitive Accumulation in Marxism, Historical or Trans-historical Separation from Means of Production?" *The Commoner* (March), available at: www.thecommoner.org.

7   Neoliberal primitive accumulation

Jim Glassman

The so-called primitive accumulation is no longer primitive. As much recent scholarship has recognized, the founding events that Marx saw as enabling capitalist accumulation proper (i.e., the process of expanded reproduction) are not just *pre*conditions of capitalism but ongoing *conditions* of its existence (DeAngelis 1999; Glassman 2006). Moreover, primitive accumulation is itself subject to expanded reproduction, taking on new forms and developing in new locations (e.g., privatization of state enterprises in highly industrialized countries), thus leading David Harvey to update the concept under the heading of "accumulation by dispossession" (Harvey 2003).

The chapters in this section contribute to our understanding of contemporary primitive accumulation in two important ways. First, they build on the developing literature that addresses what Neil Brenner and Nik Theodore call "actually existing neoliberalism" (Brenner and Theodore 2002) – neoliberalism in its varied, protean, real-world forms rather than in its thinly propagandistic self-descriptions. In this, they help us to discern some of what might demarcate the specifically neoliberal dimensions of contemporary primitive accumulation. Second, the chapters in this section focus on specific environmental dimensions – and contradictions – of primitive accumulation that help us see some potential barriers of the neoliberal project. While the main barrier to capitalist development that Marx saw as being unleashed by original accumulation was the development of a working class with nothing to lose but the chains enslaving it to capital, the possibility of environmental barriers to reproduction of capitalist relations has become an important contemporary reality (O'Connor 1988). But there is no simple, unified "nature" to pose that barrier, any more than there has turned out to be a simple, unified working class, so careful investigation of specific environmental tensions and contradictions in the neoliberal project is an important task, to which these chapters contribute.

Actually existing neoliberalism belies the simplistic rhetoric of free markets and free trade that form the core of the preferred neoliberal self-image. For example, critical scholars have long recognized recent "free trade agreements," such as the North American Free Trade Agreement (NAFTA), to be forms of *re*-regulation rather than deregulation, involving new (not necessarily reduced) roles for states in the governance of trade. Eric Swyngedouw likewise uses this term to specify what is being transformed in neoliberal regimes of urban water governance. Rather than receding, states come to play new roles in managing the dispossession that occurs as formerly public utilities are privatized and new institutional arrangements are put in place to ensure the profitability of private sector investors.

James McCarthy shows that in the case of NAFTA such "state intervention" in the economic process extends to constituting specific forms of novel and counter-intuitive property rights – under the heading of "regulatory takings" –in order to facilitate private accumulation from what might otherwise be legitimately deemed public resources. This particular maneuver thickens the plot outlined by Polanyi when he first identified land as a "fictitious commodity." Capitalist states now have as part of their institutional responsibility the production of resources as private commodities, requiring compensation to their owners for any potential (rather than actual) uses that are prevented by public use.

While such neoliberal maneuvers are not consistent with an image of neoliberalism as rolling back the state, they might seem more consistent with the view that neoliberalism centers around the institutionalization of property rights regimes favoring individual investors. Yet as Becky Mansfield shows, neoliberal practice is flexible on this point as well: the privatization of North Pacific fisheries has been accomplished through quota that can be held as readily by collectives as by individual investors.

Does any of this, then, allow for generalizations as to what might constitute a specifically neoliberal process of primitive accumulation? Or is neoliberalism infinitely flexible and opportunistic in the forms it can take. While neoliberalism certainly does seem to center in many instances around allocation of specific – if varying – forms of property rights, as noted in several of these chapters, another aspect of neoliberalism seems worth emphasizing. Harvey, following the arguments of Gérard Duménil, Dominique Lévy, and Giovanni Arrighi, has characterized neoliberalism as the class project of the most powerful and mobile capitalists, who use their geographical leverage– frequently through financialization – to roll back redistributive state projects historically won through labor struggles, thus distributing wealth and income upward (Harvey 2005; Duménil and Lévy 2004; Arrighi 1994). While Harvey strangely characterizes Fernand Braudel's theoretical perspective as "inappropriate to our contemporary world" (Harvey 2006: 80), we might take a page from Braudel here, when he notes that the most powerful capitalists are not wed to a specific industry, sector, or location but readily move their investments around the board, leaving workers and lesser capitalists to deal with the most severe cyclical and structural economic difficulties (Braudel 1979, vol. II: 432–3).

Could this be what we witness when in Guyana, as Gavin Bridge notes, absentee landholders buy up mining permits, waiting for actual mining investors to rent the land, thus using "gifts of nature as sites of circulation"? And might the Individual Transferable Quota system for halibut and sable-fish, discussed by Mansfield, allow for the same kind of leveraging of fishing industry profit by financial investors? Could this also be comparable to what we witness in Montana, where as Paul Robbins and April Luginbuhl note out-of-state investors buy up ranches for recreational game hunting at the expense of local hunting groups? And could it also be a way to interpret the movement, discussed by Swyngedouw, of a small number of French investors into positions of oligopolistic dominance in the global water market? (Braudel notes that dominant capitalists avoid free markets and competition where possible, using their power and mobility to generate monopolies and ensure higher profits.)

So perhaps neoliberalism is not a project for rolling back the state, deregulating the economy, privatizing enterprises, or even implementing private property regimes *per se*, but is rather a class practice of the most powerful, geographically mobile capitalists, who use both state rollback and state "roll out" (Peck and Tickell 2002), deregulation and re-regulation, privatization and nationalization (cf. the Mexican banking crisis of the early 1980s), and varied property regimes quite opportunistically. Neoliberal accumulation by dispossession, then, would be a sort of guerrilla war of the most powerful investors against all the rest – including even many other less powerful capitalists and business groups.

But what are the environmental consequences and limits of such a guer-rilla war? As all of these chapters illustrate, there is no predictable outcome, yet in some contexts environmental developments have short-circuited or marred the neoliberal agenda. As Robbins and Luginbuhl note, the rising incidence of Chronic Wasting Disease (CWD), accompanied by unpredicted resistance to privatization by groups of hunters, turned back attempts by larger ranchers (backed by neoliberal economic arguments) to affect insti-tutional enclosure of game such as elk. The outbreaks of malaria that Bridge notes do not seem to have as yet curtailed mining projects, but they mark one of the untoward consequences of the enclosure and development of Guyanese land and vertical territory as a fictitious commodity. And the dispossession of water Swyngedouw describes led infamously to a cholera outbreak in South Africa, delegitimizing privatization and helping spur popular struggles over water (CBC 2004).

Of course, as several of these chapters note, many types of environmental management problems are not at all specific to neoliberalism but inhere in the broader contradictions of developing natural resources as fictitious commodities. But might neoliberal primitive accumulation, as a highly mobile form of class war that demands the fungibility of all being (and thus its existence as one or another form of property), generate particularly intense contradictions between natural processes and the accumulation of capital in its most general form, as McCarthy suggests? Or will neoliberal capitalism instead develop its own resolutions to these contradictions? (As

Harvey notes, the neoliberal era is marked by both increased financialization of the economy and the slowest growth of any post-World War II period. Is this an unplanned move towards the "dematerialization" of the economy that some environmentalists have favored?) Whatever may be the case, it is certain that it will require more studies of actually existing neoliberalism, such as those conducted here, in order to determine whether or not neoliberal primitive accumulation, in its varied forms, will be consistent with the reproduction of capitalist society or instead produce its own gravediggers.

References

Arrighi, G. (1994) *The Long Twentieth Century: Money, Power, and the Origins of Our Times*, London and New York: Verso.

Braudel, F. (1979) *The Wheels of Commerce: Capitalism, 15th–18th Century*, vol. II, trans. Siân Reynolds, New York: Harper & Row.

Brenner, N. and Theodore, N. (2002) "Cities and the Geographies of 'Actually Existing Neoliberalism,'" *Antipode* 34(3): 349–79.

Canadian Broadcasting Company (CBC) (2004) "The Fifth Estate: Dead in the Water," broadcast March 31. Transcript available at: www.cbc.ca/fifth/deadinthewater.

DeAngelis, D. (1999) "Marx's Theory of Primitive Accumulation: A Suggested Reinterpretation," available at: www.homepages.uel.ac.uk/M.DeAngelis/PRIMACCA.htm.

Duménil, G. and Lévy, D. (2004) *Capital Resurgent: Roots of the Neoliberal Revolution*, trans. Derek Jeffers, Cambridge, MA: Harvard University Press.

Glassman, J. (2006) "Primitive Accumulation, Accumulation by Dispossession, Accumulation by Extra-Economic Means," *Progress in Human Geography* 30(5): 608–25.

Harvey, D. (2003) *The New Imperialism*, Oxford: Oxford University Press.

—— (2005) *A Brief History of Neoliberalism*, Oxford: Oxford University Press.

—— (2006) *Spaces of Global Capitalism: Towards a Theory of Uneven Geographical Development*, London and New York: Verso.

O'Conner, J. (1988) "Capitalism, Nature, Socialism: A Theoretical Introduction," *Capitalism, Nature, Socialism* 1: 11–38.

Peck, J. and Tickell, A. (2002) "Neoliberalizing Space," *Antipode* 34(3): 380–404.

Part II
Commodification and marketization

8 Neoliberalizing nature?

Market environmentalism in water supply in England and Wales

Karen Bakker

Introduction

The 1989 privatization of the water supply sector in England and Wales is a much-cited model of "market environmentalism": the application of market institutions to natural resource management as a means of reconciling goals of efficiency and environmental conservation. With the privatization of the water supply industry in 1989, ownership passed from nationalized monopolies to private companies, listed on the London Stock Exchange. The controversy over privatization has often obscured the fact that a much broader transformation of water supply management in England and Wales has occurred over the past three decades. Demand management has been increasingly prioritized over supply-side management strategies (such as dams and other large-scale hydraulic infrastructure). Economists and environmental scientists have supplemented (and to some extent displaced) engineers in managerial positions. Water is no longer perceived to be universally abundant: "areas of water scarcity" have been enshrined in legislation. Efficiency and cost-reflectiveness are prioritized over social equity in water pricing; national cross-subsidies have disappeared and regional cross-subsidies dwindled. Consumers are characterized as "customers" rather than "citizens," and their means of participation in policy debates has changed significantly, although their influence has not necessarily increased. Environmental and drinking water quality have improved; according to the environmental regulator of the industry, river water quality in Britain is at its highest level since the Industrial Revolution (Bakker 2004).

The "great transformation" in water supply management in England and Wales is thus multi-faceted. This chapter explores the nature of this transformation, and seeks to answer three questions. First, what have been the impacts of this substantial re-regulation of water in England and Wales, for consumers, workers, companies, and the environment? Second, what is the analytical utility of the term "neoliberalism" in describing these changes; and how does this term need to be refined in order to adequately describe the nature of the re-regulatory process? Third, can the project of water supply privatization and re-regulation be categorized as a success or failure, and on what grounds?

Market environmentalism

The arguments of the proponents of neoliberal resource management are perhaps best captured by the term "market environmentalism": a mode of resource regulation which promises both economic and environmental ends via market means (Anderson and Leal 2001). Some political economists have framed market environmentalism as a form of "green imperialism" – whereby specific instances of environmental degradation (an inevitable if unintended by-product of capital accumulation) are mobilized as opportunities for continued profit. Others focus on the political economic dimensions of privatization; Harvey, for example, characterizes privatization of water supply as one example of "accumulation by dispossession" – the enclosure of public assets by private interests, for profit, resulting in greater social inequity (Harvey 2003). Still other political economic approaches have focused on the dynamics of resource regulation, seeking to articulate specific neoliberal projects with analysis of generalized transformations in modes of political economic governance (Gandy 1997). Drawing on Foucauldian governmentality theory, attention has also been paid to "neoliberalism" as a project of environmental governance (McCarthy and Prudham 2004). From this perspective, neoliberalism is understood to be more than merely a political economic project with impacts on the environment; rather, neoliberalism is conceptualized as being constituted by (and of) processes of socio-environmental change.

Concepts of privatization, commercialization, marketization and commodification figure centrally in much of this work. Yet, as Noel Castree observes in his review of recent work on the commodification of nature, these concepts are often conflated (Castree 2003). Privatization is often assumed to entail commercialization and commodification, to the extent that the terms are at times used interchangeably. Moreover, much of this work is, if only implicitly, normative: commodification, markets, and private sector actors are understood to be pernicious; often, the impacts of neoliberalism upon the environment are assumed to be largely negative. As Castree observes, analytical imprecision and the failure to make explicit the normative bases of our arguments have significant consequences: occluding processes of commodification in some instances; and undermining the progressive potential of critical scholarship, in others. This difficulty is compounded by an analytical focus on *neoliberalism* as a hegemonic, singular project – which encourages excessively generalist categorizations of neoliberalism in some cases, and unreflexively concrete and contingent analyses of local neoliberal projects in others (Peck and Tickell 2002).

Privatization, commercialization, and commodification

The analysis presented in this chapter attempts to clarify some conflations and question some of the assumptions of this literature. In particular,

I attempt to show that privatization does not necessarily involve deregulation, but is rather a process of selective re-regulation which can have positive, as well as negative environmental impacts. In order to make this claim, greater analytical precision is required with respect to the term "resource re-regulation," which usually involves three interrelated processes: privatization, commercialization, and commodification. *Privatization* entails a change of ownership, or a handover of management from the public to the private sector. *Commercialization* entails changes in resource management practices which introduce commercial principles (such as efficiency), methods (such as cost-benefit assessment), and objectives (such as profit-maximization). Privatization thus entails organizational change, in distinction from commercialization, which entails institutional change (in the sociological sense of rules, norms, and customs). Privatization and commercialization (although often inter-related) must be understood as distinct processes. Privatization can occur without full commercialization, as is the case with many water companies in developing countries – where private, for-profit companies operate tariff structures which price water on a below-marginal cost basis to poorer customers. Commercialization can be initiated prior to privatization, or while ownership is retained in the public sector. For example, many publicly owned utilities in the OECD employ rising block tariffs and universal metering to price water at full cost.

From a neoliberal perspective, neither privatization nor commercialization will ensure the conversion of resources into commodities. *Commodification* entails the creation of an economic good, through the application of mechanisms to appropriate and standardize a class of goods or services, enabling them to be sold at a price determined through market exchange. Commodification and commercialization are related, but analytically distinct: the latter entails changes in resource management institutions, a necessary but insufficient condition for the former, which involves the conversion of a resource into an economic good – by no means a straightforward process, as neoclassical economists recognize when referring to the multiple "market failures" which characterize resources such as water supply. Yet, from a neoclassical perspective, the conversion of a resource into an economic good is necessary if water is to be more efficiently managed. In other words, privatization and commercialization are necessary, but insufficient conditions for optimal water management. Understanding the distinctions drawn by the above typology is important. Failing to disaggregate privatization, commercialization, and commodification runs the risk of obscuring a critically important dimension of neoliberal projects, which weakens our ability to understand how neoliberalization evolves, and why neoliberal projects may sometimes falter.

Attempts to neoliberalize resource management are particularly fraught with difficulty in the case of what Benton terms "eco-regulatory" production, which simultaneously circumscribes, transforms and adapts to "nature" as "resource" (Foster 2000). This is in part because commodification is a politically contentious process, insofar as it must "play out upon, as well as produce, a diverse ecological landscape" (Robertson 2000: 466) invested with divergent political and economic interests. In an attempt to counter these contradictions, re-regulation of resources occurs as public and private actors respond in a variety of creative, and constantly evolving ways: capital seeking profit; the state seeking to develop a mutually supportive relationship between capital accumulation and regulation, enabling economic growth and creating the conditions for political stability – an example of Polanyi's "double movement," which may lead to significant re-regulation of specific sectors. Neoliberalization may not, in other words, imply de-regulation, but rather selective re-regulation; indeed, as subsequent sections of the chapter explore, the water supply sector in England and Wales has been substantially re-regulated since its privatization in 1989.

The "great transformation" in water supply in England and Wales

The collapse of the "state hydraulic" paradigm

Throughout much of the twentieth century, the water supply industry in England and Wales was run on a monopolistic basis, and regulated as a public service, with the majority of infrastructure owned by governments (municipal and then national). Drinking water was supplied with the goal of universal provision. Water pricing was based on a concept of "social equity": household supply was not metered, and bills were linked to property value, supported through cross-subsidies between consumers, and in some instances between regions and level of governments. Potable water was a key concern for the developers of water supply networks, who were keenly aware of the links between polluted water and the cholera and typhoid epidemics that ravaged nineteenth-century cities. Water planners focused on developing new water sources such as reservoirs, pursuing a supply-led strategy to anticipate increasing water demands stemming from economic and population growth.

Significant problems emerged with respect to this "state hydraulic" paradigm, which lent support to calls for the progressive commercialization of the industry according to a "market environmentalist" logic. Crucially, under-investment in infrastructure (to minimize public sector borrowing for macro-economic reasons, and to maintain low water bills for political reasons), and sustained industrial water pollution contributed to the continued decline of river and tap water quality in Britain for decades (Kinnersley 1994; Summerton 1998). The much-lauded integration of water supply and regulatory functions in basin-wide Regional Water Authorities according to the principle of Integrated River Basin Management had the unforeseen side-effect of discouraging enforcement of water quality regulation (particularly sewage works), further aggravating environmental degradation. The decision of the European Union to prosecute Britain for non-compliance in the mid-1980s was politically decisive (Hassan 1998); increased capital investment to meet European water quality standards was unavoidable, and estimated investment requirements for the following decade ranged from £24 to £30 billion (1989 prices), amounts which the Conservative government was unwilling to spend.

Re-regulation: the consolidation of market environmentalism

The subsequent decision to privatize the water industry was the apogee of the Conservative government's privatization programme. Privatization consolidated the commercialization of the water supply industry through the introduction of market-simulating regulatory mechanisms such as cost-benefit analysis into both economic and environmental regulation. Little over a decade after privatization, labour levels have been dramatically reduced, collective bargaining mechanisms dismantled, and out-sourcing "non-core" functions has significantly changed labour relations and practices in the industry (O'Connell-Davidson 1993). Investment levels have increased, with companies spending £31 billion from 1990 to 2000; investment over the period from 1991 to 1996 was twice levels prior to 1989 (Kinnersley 1998). In pricing, economic equity is prioritized over social equity (Bakker 2001). In economic regulation, efficiency is prioritized, although the increase in efficiency of water supply management is disputed. Water companies have been consistently profitable, although rates of profit have dropped as the price-cap regulatory regime has been progressively tightened, with a corresponding drop in share prices, albeit amidst controversy over "fat cat" salaries and the extent to which the increase in consumers' bills above costs of doing business (rather than increases in efficiency) is a contributing factor to profitability (Saal and Parker 2001; Shaoul 1997).

Water supply system management practices have evolved significantly: rather than engineering-driven approaches prioritizing redundancy and interconnection in the storage and distribution networks (and hence security of supply), economics-driven approaches prioritizing economically efficient management of the network and demand management (and hence on cost minimization for given output) are increasingly central to water resource management policies. This shift stems in part from growing concerns about the impacts of climate change on water resource security, particularly in southern England, and an increasingly dominant discursive depiction of water as a scarce resource (noteworthy in such a "wet" country) – which have recently been enshrined in U.K. legislation with the designation of official "Areas of Water Scarcity."

Another driver is the prioritization of environmental concerns. Environmental issues have been formally integrated into water resources planning, and the water industry has to some degree reinvented itself as an "environmental services" industry. The creation of a separate environmental regulator has elevated the environment to the status of "legitimate user" – with visibility and clout – within the regulatory framework. Much greater emphasis is placed on aesthetics, amenity value of landscape, and the value of "natural landscapes" – incorporated in environmental economic valuation, instrumentalized through changes to pricing of water abstraction, and valorized through river restoration projects. Water quality and environmental expenditure are key drivers of capital expenditure programmes in the industry; with an estimated expenditure of between £8 and £8.5 billion on water quality between 2000 and 2005 on water quality, much of this directed towards improving the quality of discharges from sewage treatment works and ending the practice of direct disposal of sewage to sea or waterways through combined sewer overflows. Partly as a result, chemical and biological river water quality has improved, although compliance with river water quality objectives set by the government had reached only 82 percent in 1999, a level which is viewed by the government as unsatisfactory. Drinking water quality has also improved significantly. Much of this improvement is driven by increasingly comprehensive European Union water quality legislation governing beaches and bathing waters, drinking water quality, and environmental quality of both surface and groundwater (Walker 1983; Buller 1996; Kallis and Butler 2001; Kaika 2003). Water companies in England and Wales are to a much greater extent guided and constrained by environmental regulations than they were three decades ago. So too are managers, whose performance-based incentive schemes now routinely incorporate environmental performance criteria (Hopkinson *et al.* 2000), backed up by the threat of prosecution or "naming and shaming" by the environmental regulator through high-profile public reporting.

The increasing dominance of environmental concerns is characteristic of the shift in relative influence of different stakeholders under market environmentalism – with labour unions sidelined, and consumers' interests balanced with, or trumped by environmental concerns. Environmental externalities are addressed within the water policy framework, and backed up in most instances by legal obligations. In contrast, social externalities are now, to a greater degree than in the past, excluded from the water policy framework (Bakker 2004). These shifting power geometries are most clearly observed in the formal structure of regulation: whereas the environmental regulator is a separate well-funded entity, the regulatory body responsible for consumers has, until recently, operated under the aegis of the economic regulator, the Office of Water Services (Ofwat), with a highly constrained role. A significant proportion of the increases in domestic water users' bills post-privatization has been due to environmental expenditure, producing clear gains for the environment in some cases, but at the apparent cost of consumers; hence the frequent disagreements between environmental groups and consumers groups over water policy, particularly given the highly controversial impacts of water debt and "water poverty" on public health (Drakeford 1997).

The drive for commodification

As explored above, re-regulation consolidated and deepened the progressive commercialization of the water supply industry. Yet successive governments (both Conservative and Labour), as well as the economic regulator (Ofwat), envisioned a further step: the gradual commodification of water. This entailed a two-pronged strategy: (1) introducing accurate prices and competition as a means of enabling market exchange through introducing universal metering, environmental economic valuation, and (2) introducing direct competition through the integration of water supply networks.

To argue that water is not a commodity, despite the fact that it has a price and is delivered by private companies, may seem at first disingenuous. But water clearly did not fit the neoclassical definition of a commodity: a standardized good or service, with interchangeable units, sold at a price determined through market exchange. Post-privatization, in England and Wales, water was not standardized: water quality varied considerably (chemically and biologically) between catchments; and water supply networks were not integrated even within company supply areas. Water companies remain vertically integrated monopolies, responsible for everything from raw water abstraction through to delivery to the customer. With no national, and few truly regional water grids in the U.K., water was not traded in bulk, and water transfers between companies were limited in volume.

The principles underlying water supply thus remained that of the state hydraulic era, and networked water supply in England and Wales remained (from a neoliberal perspective) a "quasi-commodity," or only partially commodified. This was not unexpected; neoclassical economists use the term "market failure" to describe instances where goods fail to meet the necessary criteria for commodification. They identify two important "market failures" with respect to networked water supply: "natural monopoly" (supply by one firm entails lower costs than supply by more than one firm), and "externalities" (costs or benefits arising from water production not accounted for in the price mechanism, which thus do not accrue to the producer).

Accordingly, two barriers to commodification were the focus of Department of Environment policy-making post-privatization: the absence of competition; and the lack of market-based pricing mechanisms (implying a continuation of widespread cross-subsidies, failure to incorporate externalities in water pricing, and an absence of accurate price-signals). The resulting difficulty, from the perspective of Ofwat and the government, was that the production of water would be less than optimal; in the absence of competition and adequate price-signaling mechanisms, the market will not function as an efficiency-maximizing institution for the allocation of resources. The solution, as explored in the following section, was to introduce market pricing and market competition: processes that have proven to be fraught with difficulty.

Competition stymied

Networked water supply, as a naturalized monopoly, poses a particularly intractable challenge to market environmentalism. Under the state hydraulic paradigm, protecting consumers against the abuse of monopoly powers was accomplished through direct government vetting of prices and investment programmes, and, in many instances, public ownership and management of infrastructure. Under market environmentalism, competition is assumed to be a better mechanism than command-and-control regulation, legislation, or moral suasion. As two of the best-known proponents of free market environmentalism in the U.S. assert: "good resource stewardship depends on how well social institutions harness self-interest through individual incentives" (Anderson and Leal 2001: 5). Self-interest, in the case of water supply companies, equates to profit, to be harnessed via competition, thereby encouraging innovation and – as expected by the architects of economic regulation for privatized utilities in Britain – driving down consumer prices.

The commitment to introducing competition is reflected in the statutory duty of the economic regulator, Ofwat, to facilitate competition. Given the lack of a national grid and recognized "market failures," the post-privatization regulatory framework focused on surrogate competition – through a system of "comparative" or "yardstick" competition administered by an economic regulator – and competition for corporate control (mergers and takeovers). Under yardstick or comparative competition (and unlike American-style rate-of-return regulation, in which dividends are capped) utilities' maximum price increases are capped based on regulatory comparisons of company performance and efficiency. In theory, the incentive for a water company to increase efficiency arises from the fact that companies can increase profit by increasing efficiency, thereby retaining expenditure in addition to the revenue implicitly allowed by their price cap (Littlechild 1988). Comparative competition thus relies on a set of benchmarks, which are in theory a function of all firms' performance, thus diminishing (if not eliminating) the scope for strategic behaviour (such as inflation of cost projections) on the part of the private company. With price caps set in advance, competition amongst companies occurs relative to the efficiency "yardsticks" calculated by the regulator, backed up by the threat of takeover in case of poor performance. The profit motive is thus, in theory, harnessed by comparative competition-driven price cap regulation to drive efficiency gains and reduce costs.

The failure to control prices and profits is one reason why comparative competition had begun, by the late 1990s, to be viewed by many within the industry as a "pale and sickly relative of market competition" (Summerton 2001: 23). Another, more fundamental issue also troubled the economic regulator, Sir Ian Byatt: the difficulty in comparing company performance. In order to calculate the price caps, the regulator employs econometric models and detailed assessments of individual company performance to identify potential reductions in operating, capital maintenance, and capital enhancement expenditure (Ofwat 1998a). Comparative competition thus entails the calculation of potential efficiency gains not only through reference to individual companies, but also through the relative ranking of company performance; an information-intensive and costly exercise – the budget for the economic regulator alone was just under US$25 million in 2002–3, in addition to the additional US$25 million spent by companies on reporting and auditing requirements. Despite the scale of the regulatory exercise, technical difficulties arose in comparing companies following privatization, notably variations in environmental conditions (Hopkinson *et al.* 2000).

A second problem arose with respect to the contradiction between comparative and corporate competition: the threat of takeover had worked so well that mergers reduced the original 39 companies down to 22 by 2004. As a result, the UK's Competition Commission has prohibited recent mergers, citing the need to retain a sufficient number of comparators in order for the economic regulator to carry out robust comparative competition. Comparative competition thus contains a seemingly intractable dilemma: the preservation of a sufficient number of distinct water suppliers is necessary to underpin comparative competition; but this reduces the threat of takeover as a "spur to efficiency."

Given the limitations on comparative competition, much government and regulatory effort was expended post-privatization in an attempt to introduce direct competition, through various arrangements designed to facilitate "common carriage" of bulk water through supply networks – allowing customers to choose water suppliers much as they would choose a provider of telephony services. Attempts to introduce direct competition were stymied by high costs, technical barriers, and concerns over public health. Plans to introduce a market in tradable abstraction permits were also dropped, largely due to concerns over public and environmental health.

In summary, the government was forced to retreat on plans for direct competition by the late 1990s. As a result, direct competition in the water supply sector remains limited in comparison with the expectations of the architects of privatization, and in comparison to other privatized utility industries in the U.K. The regulatory system continued to rely largely on comparative competition through "yardstick" regulation, which undermined the original goal of liberalizing the water sector post-privatization.

Metering and the repoliticization of pricing

The second key area of retrenchment of the market environmentalist project arose with respect to water pricing. In addition to properly valuing water in order to establish its true costs, commodification requires a mechanism for communicating price signals to consumers. Accordingly, some form of volumetric metering is required. At the time of privatization, 100 per cent penetration of meters into domestic properties was envisaged; full metering was originally required of the water supply industry by the year 2000. Yet by the late 1990s, the government had quietly dropped the obligation on the part of water companies to meter all customers. Although over 99 per cent of the population of England and Wales is connected to a water supply network, by 2000 domestic metering penetration levels had not yet reached 20 per cent, one of the lowest levels in the OECD (Day 2003). As water prices rose significantly above the rate of inflation post-privatization, an increasing number of consumers failed to pay bills on time, and disconnections for non-payment increased. Together with the introduction of pre-payment meters (largely into the homes of low-income consumers) this came to be associated in public discourse with "water poverty" (Bakker 2001).

This marked a dramatic retreat from the original pricing model envisioned by the architects of privatization. Under market environmentalism, the justification for full cost pricing is its supposed effect on consumer and environmental welfare – defined in economic terms. A minimization of prices for a given level of service is predicted to be the result of efficiency gains in water services provision. In legislation and in practice, it should be noted, the minimization of prices is not an explicit goal; the promotion of efficiency is an explicit goal, from which the minimization of prices for a given level of output is expected to result.

The architects of privatization assumed that these duties would be complementary: customers would benefit if efficient companies remained financially viable. The regulatory framework, however, contains no explicit mechanism for addressing the question of the acceptable level of these prices. Accordingly, balancing the need to generate stable, sufficiently high levels of return to satisfy investors on the one hand, and politically acceptable rates of return, on the other, has been accomplished through political intervention.

With the election of Tony Blair's Labour government to power in 1997, measures to protect vulnerable consumers were instituted; price caps were reduced to 0 per cent above inflation at the 1999 Periodic Review (Bakker 2001). Water companies' rates of profit declined from over 10 per cent to 6 per cent (pre-tax). In contrast to other privatized utility sectors (such as telecommunications), where privatization and price cap regulation have led to increasing competition and reductions in prices, the water supply sector proved to be the most problematic for the British model of privatized utility regulation. As a natural monopoly essential for public health and discursively constructed as an emblem of inclusionary citizenship, water supply access and affordability remain highly politicized. As stated by the Department of the Environment in 1985, one of the key objectives of the privatization programme was to "free enterprise from state controls." However, as a senior economist with the Office of Water Services (the economic regulator) recently noted: "water remains a matter of public policy . . . political and social concerns are alive and well as key influences on the pattern of water prices" (Day 2003: 41) – precisely what the architects of privatization had hoped to avoid.

Conclusions: uncooperative commodities

Post-privatization, both the government and the economic regulator were intent on fully commodifying water. The conversion of water into an economic good required the introduction of true competition (via integrated, trans-watershed infrastructure networks), and cost-reflective pricing (requiring new environmental valuation techniques, and technologies such as meters in order to convey price signals). After a decade of experimentation, both of these initiatives have been substantially retrenched. Market environmentalism in water supply in England can thus be characterized as a case of successful privatization, retrenched commercialization, and failed commodification.

As explored above, this failure to commodify water is in large part due to water's geography: a life-giving, continually circulating, scale-linking resource whose biophysical, spatial, and socio-cultural characteristics render it particularly resistant to commodification. The ecological and possible public health consequences of network integration of a flow resource effectively prevented the introduction of direct competition; the introduction of accurate prices was stymied by political resistance to metering and price increases due to "water poverty," and the difficulty of incorporating robust environmental economic valuation techniques. These contradictions could not be resolved within the post-privatization regulatory framework, leading to substantial re-regulation of the water industry. Moreover, intense political debate about water's identity – as entitlement for citizens, or as commodity for customers – further destabilized the market environmentalist project. Hence, the government retreated from water valuation and liberalization, scaling back on plans to introduce direct competition and trading of abstraction licences, dropping the requirement for universal metering, and re-inserting social considerations, particularly for vulnerable consumers, into the pricing framework.

In summary, the application of market mechanisms to water supply management is much more limited than had been expected in England and Wales. The biophysical, environmental, health, and political aspects of water supply undermined attempts to commodify water; this failure was a critical driver in the re-regulation of the water supply industry, and in the overall trend towards improvement in environmental and drinking water quality, a finding which underpins my closing argument – that neoliberalization is implicated in processes of re-regulation which rescript the entitlements of both humans and non-humans, with outcomes that are not necessarily negative for what we conventionally delimit as the "environment."

My argument is not that neoliberalism is causally related to improvements in environmental quality. Rather, my point is that neoliberalization is constituted by (and constitutive of) processes of re-regulation that may result in improvements in environmental quality. The difference between these two arguments is subtle, but important. The former asserts a causal relationship; the latter, in contrast, cautions analysts not to jump to conclusions about causality, and in particular, not to assume that environmental quality can only decline in the context of neoliberalization of resource regulation.

In turn, this enables a focus on the progressive possibilities opened up within the current international trend towards market environmentalist resource management. This is particularly relevant to the case of water. Some of the great gains in human welfare during the twentieth century associated with the "state hydraulic paradigm" were made at the expense of the environment – with the state temporarily devolving costs onto the environment in what might be termed an "ecological fix" (Bakker 2004), rationally administering massive environmental degradation and systematic under-provision of environmental goods. Attitudes toward the state become more ambivalent (and the conflation of "state" with "public" interest more obviously erroneous) when one factors the environment into the redistributive equation. In the case of market environmentalism in England and Wales, improved water quality and increased protection against domestic disconnections are respective examples of progressive environmental and social aspects of this transition. Yet the balance of cost allocation has shifted; whereas the social costs of water production were previously externalized from the sphere of the politicized citizen and borne by the environment, the environmental costs of water production are now (to a greater degree) externalized from the sphere of capitalized environment and borne by consumers. Of course, the distinction between "environmental" and "social" costs is a constantly shifting and unstable divide; as David Harvey has repeatedly pointed out, ecological projects are always sociopolitical projects (and vice versa). The task for the analyst, as attempted in this chapter, is to identify how our collective commitment to socioenvironmental justice has been altered as a result.

Acknowledgements

A longer version of this chapter appeared as Bakker, K. (2005) "Neoliberalizing Nature? Market Environmentalism in Water Supply in England and Wales," *Annals of the Association of American Geographers* 95(3): 542–65.

References

Anderson, T. and Leal, D. (2001) *Free Market Environmentalism*, New York: Palgrave.
Bakker, K. (2004) *An Uncooperative Commodity: Privatising Water in England and Wales*, Oxford: Oxford University Press.
Ballance, T. and Taylor, A. (2005) *Competition and Economic Regulation in Water: The Future of the European Water Industry*, London: IWA Publishing.
Castree, N. (2003) "Commodifying What Nature?" *Progress in Human Geography* 27(3): 273–97.

Drakeford, M. (1997) "The Poverty of Privatization: Poorest Customers of the Privatized Gas, Water and Electricity Industries," *Critical Social Policy* 17: 115–32.

Ernst, J. (1994) *Whose Utility? The Social Impact of Public Utility Privatization and Regulation in Britain*, Milton Keynes: Open University Press.

Foster, C. D. (1992) *Privatization, Public Ownership and the Regulation of Natural Monopoly*, Oxford: Blackwell.

Foster, J. (2000) *Marx's Ecology: Materialism and Nature,* New York: Monthly Review Press.

Gandy, M. (1997) "The Making of a Regulatory Crisis: Restructuring New York City's Water Supply," *Transactions of the Institute of British Geographers* 22: 338–58.

Goubert, J. P. (1986) *The Conquest of Water*. London: Polity.

Harvey, D. (2003) *The New Imperialism*. New York: Oxford University Press.

Hassan, J. (1998) *A History of Water in Modern England and Wales*, Manchester: Manchester University Press.

Hopkinson, P. *et al.* (2000) "Environmental Performance Evaluation in the Water Industry of England and Wales," *Journal of Environmental Planning and Management* 43(6): 873–95.

Kinnersley, D. (1994) *Coming Clean: The Politics of Water and the Environment*, London: Penguin.

Lerner, W. (2003) "Neoliberalism?" *Progress in Human Geography* 21: 509–12.

Littlechild, S. (1988) "Economic Regulation of Privatised Water Authorities and Some Further Reflections," *Oxford Review of Economic Policy* 4(2): 40–68.

McCarthy, J. and Prudham, S. (2004) "Neoliberal Nature and the Nature of Neoliberalism," *Geoforum* 35(3): 275–83.

McMahon, P. and Postle, M. (2000) "Environmental Valuation and Water Resources Planning in England and Wales," *Water Policy* 2: 397–421.

O'Connell-Davidson, J. (1993) *Privatization and Employment Relations: The Case of the Water Industry*, London: Mansell.

Peck, J. and Tickell, A. (2002) "Neoliberalizing Space," *Antipode* 34(3): 380–404.

Robertson, M. (2000) "No net loss: Wetland restoration and the incomplete capitalization of nature," *Antipode* 32(4): 462–93.

Saal, D. and Parker, D. (2001) "Productivity and Price Performance in the Privatized Water and Sewerage Companies of England and Wales," *Journal of Regulatory Economics* 20(1): 61–90.

Saunders, P. and Harris, C. (1994) *Privatisation and Popular Capitalism*, Buckingham: Open University Press.

Shaoul, J. (1997) "A Critical Financial Analysis of the Performance of Privatized Industry: The Case of the UK Water Industry," *Critical Perspectives on Accounting* 8: 479–505.

Strang, V. (2004) *The Meaning of Water*, Oxford: Berg.

Summerton, N. (1998) "The British Way in Water," *Water Policy* 1: 45–65.

Swyngedouw, E. (1999) "Modernity and Hybridity: Nature, Regeneracionismo, and the Production of the Spanish Waterscape, 1890–1930," *Annals of the Association of American Geographers* 89(3): 443–65.

Vickers, J. (1997) "Regulation, Competition and the Structure of Prices," *Oxford Review of Economic Policy* 13(1): 15–26.

Winpenny, J. (1994) *Managing Water as an Economic Resource*, London: Routledge.

9 The neoliberalization of ecosystem services

Wetland mitigation banking and the problem of measurement

Morgan M. Robertson

Introduction

Just northwest of Aurora, Illinois, on the outskirts of Chicago, lie one hundred and twenty acres of silt loam soils, planted in the latest variety of feed corn. Nailed to a machine-shed, a large "For Sale" sign indicates that this particular piece of real estate is about to be thrown into the circuit of capital once again, but this time there will be a difference: the prospective new owner, a residential development firm, is not planning to plant corn or soybeans. Instead, they will produce and sell *ecosystem services*, by restoring the site to its presettlement wetland condition, and establishing a *commercial wetland mitigation bank*. By entering into a complex agreement with federal and county regulatory agencies, the banking firm will sell "wetland credits" to individuals compelled to buy them by those same agencies. Within five years, the production and sale of ecosystem services in the farm field outside of Aurora will have grossed nearly three million dollars.

Commercial wetland mitigation banking is the product of an American environmental management policy that seeks to develop a market in privately-owned "wetland ecosystem services," such as duck habitat, flood protection and biodiversity, as a way of achieving the goals of the US Clean Water Act of 1977 (CWA). Young though it is, it is the most mature effort yet to create commodity markets in ecosystem services *per se* in the United States. While well-deserved attention has been paid to the development of markets in air pollution abatement credits (Tietenberg 1985) and in tradable harvest rights in fisheries (McEvoy 1986; St. Martin 2001), there have been very few actually-functioning markets in which the commodity is defined and measured as a holistic character of ecosystems (rather than being measured in tons of pollutant, or number of fish). And yet it is a continual refrain of neoliberal environmental economists that ecosystem services, and not just their material components, are the commodities of the future in market-based environmental policy (Daily 1997; Costanza *et al.* 1997; Heal 2000; Daily and Ellison 2002).

I will focus here on one aspect of banking which complicates the smooth neoliberal account of the process of commodifying ecosystem services: the problem of measurement. Countless studies of capitalist modernity have made it clear that standardized methods of abstraction are a basic tool in the regulation of a smoothly-running socio-economic system (*e.g.*, capital treats only with labor-power, which is an abstract quanta homogenizing diverse particular labors). But nothing has vexed the banking community so much as the task of creating abstract and generalizable *measures* of the commodity that they sell.

The story of wetland banking indicates that there may be important differences *within* neoliberal strategy that geographers are well positioned to investigate. I suggest that the massive process of codifying and commodifying the ecological relations around us is a never-concluded project of disciplining both ecosystemic relations and people as consumers of these relations.

What is commercial wetland mitigation banking?

> After all, capital is about *creative* destruction, not simply ecological degradation.
> (Castree 2002: 141)

When the 1972 Federal Water Pollution Control Act (FWPCA) was passed, it gave the US federal government what is still one of its most far-reaching powers to regulate land-use: the power to regulate dredging or dumping in wetlands (seasonally or shallowly inundated land). Section 404 of the FWPCA (reauthorized as the 1977 Clean Water Act) provides for a permitting system: if someone wishes to fill a wetland, they must apply for a permit from the Regulatory Branch of the local District of the US Army Corps of Engineers (Corps). Following consultation with the regional office of the US Environmental Protection Agency (EPA), the Corps Project Manager may then allow them to proceed, or deny them a permit altogether, or thirdly, allow them to proceed on the condition that they create or restore a certain amount of wetland to compensate for the loss of a natural wetland. This action is known as *compensatory mitigation*. Commercial wetland banking is a regulatory arrangement by which a private firm creates compensatory mitigation sites by restoring a former wetlands area to a sufficiently functional and diverse condition. People required to perform compensatory mitigation can then purchase "wetland credits" from this firm, instead of creating the wetland themselves.

Because bankers are free to compete and set any price for credits, banking promised to provide the price signals necessary for a real market in wetland services. State and federal highway agencies had been "banking" wetlands by constructing large mitigation sites before they were needed, solely for their own use, since the early 1980s. A market in ecosystem service commodities was new, however, and seemed to be perfectly in tune with the neoliberal turn in American politics.[1]

But banking was not legislated into existence in Washington, DC. In early 1991, Chicago-area earthmoving contractor Bob Terry[2] was searching

for a way to make Section 404 compliance simpler for the development community, whose biggest complaint was spatial: the Corps preferred that compensation wetlands be constructed on the same parcel of (often very expensive)
land on which the impact was occurring. Because this construction tended to
throw off budgets and timetables, and occupy land on which developers would
rather build houses, Terry was looking for a way to build wetlands *before*
they were needed, and on cheaper land. He hit on the idea of banking:

> I just said, "Maybe I'll build some big-ass wetlands somewhere, some-
> where out there, and build some really good ones, and that ought to
> make these agencies really happy." And when we met with the agencies,
> they said, "Gee, this sounds like mitigation banking that some of these
> other agencies are doing ... " And my comment was, "Well, whatever
> you want to call it, I mean, the idea sounds like a good idea, and we
> want to do it."
>
> (interview, 7/25/2001)

Terry drafted plans to restore a site to wetlands, and began a series of
meetings with local officials of the EPA and the Corps to create the first
bank instrument agreement. In the end, Terry sold hundreds of thousands
of dollars worth of wetland credits to other real estate developers holding
Section 404 permits, after which a parks district took possession of a high-
quality wetland site, and the original developer fulfilled its obligation to
commit a portion of the development to open-space use.

In outline, what happened is that a coalition of local regulators and
businesspeople formed a network based largely on acquaintance and proximity to draft the rules by which something called a "wetland credit" can be
defined and traded as a commodity. Though some were representatives of
the nation-state, they acted with only the vaguest of federal directives that
market-based tools were desirable. This was possible within the Chicago
District of the Corps because of many factors: the relative autonomy of
Corps Districts; the spatial coincidence of the regional offices of all major
federal environmental regulatory agencies; an extraordinary level of coordination between these agencies due to the fact that staffers move serially
between them; and a community of private land developers who, in a wet-
lands-dense landscape, have had to become very knowledgeable about the
Section 404 permitting system. The process of drawing ecological relations
into circuits of capital bears no resemblance to a simple fiat of capital, or to
a simple directive from the federal coordinators of a hegemonic project of
environmental governance (Henderson 1998; Robertson 2004).

Neoliberal nature

The development of banking complicates economists' streamlined narratives
about how, where and why markets in new commodities form; it continues

to be a process extended by personal relationships and geographic proximity. But how does this institutional arrangement address its commodity, and attempt to throw its arms around a concept as polysemic and unsettled as "nature"?

The fundamental neoliberal conceit concerning the environment posits that the environment is *that which is common to all of us*, the spatially-differentiated matrix of economic activity, an external presence whose dynamics affect us all. Hence, global environmental management emerges as a main trope of neo-liberal globalization: if people of all nations are to be treated as rational economic agents (which goes without saying) who differ only in our subjective preferences, then a strategy which focuses on "the global environment" (the matrix through which all such agents move) is one more way of sweeping aside troublesome institutional/collectivist obstacles to trade (Taylor 1997; Goldman 1998). This imagined general interest justifies – indeed, demands – the management of these external forces through a mechanism that is believed to guarantee the equal treatment of all: the market mechanism.

Knowledge about this external presence is assumed to be provided by ecosystem scientists, who take on a very important role in ecosystem service markets; imposing market relations on uncapitalized environmental phenomena is no easy task. Among other things, it requires techniques by which a dollar value can be placed on "environmental services," and such techniques – mixing ecological and economic principles – have proliferated over the past decade on the strength of consensus on the need to price nature. "Although ecosystem valuation is certainly difficult and fraught with uncertainties, one choice we do not have is whether or not to do it," says one prominent economist, who describes the Earth as "a very efficient, least-cost provider of human life-support services," the entire value of which is between 16 and 54 trillion dollars (Costanza *et al.* 1997: 255). As an unintended paraphrase of British Prime Minister Margaret Thatcher's famous declaration that "there is no alternative" to privatization and free-market capitalism – considered one of the great rallying-cries of neoliberal policy – Costanza's imperative places the scientific work that describes eco-systems as a bundle of services firmly within the neoliberal project. If environmental goods can be alienated and owned, the economists tell us, they will then behave as commodities behave: those environmental services that are in demand (clean air and water) will increase in supply. Similarly, once environmental harms (such as wetland destruction) have been valued and defined as property or legal obligations, payment of the social costs of these harms can be determined by bargaining between the parties involved (Coase 1960; Hockenstein *et al.* 1997; Heal 2001). In this way, the lowest-cost solution is adopted, which by definition creates the highest welfare. Despite the rather cartoonish simplicity of this account – authorless, actorless, and free of history – it is accepted as self-evident in most environmental policy discussions: "Market-based instruments are regulatory devices that shape behavior through price signals rather than explicit instructions on pollution control levels or methods. ... In a correctly functioning market, one would expect the price of a permit to equal the marginal cost of lowering emissions" (Hockenstein *et al.* 1997: 14).

The problem of measurement

In this conception, there is no room for the view of science as a shifting and constructed set of knowledges; the degree to which scientists can objectively and noncontroversially define nature as a stable external presence is the degree to which it can be rendered as a commodity and sold in markets that reveal accurate price signals. But given that the sociology of scientific knowledge cannot be wished away by the neoliberal conceit, many of the problems and contradictions of this attempt to fashion stable markets in ecosystem services concern issues of scientific measurement. For a stable system of prices to emerge, the institutional rules of the market must be clear on how the value of the commodity is to be measured. Wetland loss at the site of impact must be rendered commensurable with wetland gain at the site of banking in a regular and reliable way. In whatever way the commodity is defined, if two different developers both buy one credit of mitigation at a bank, some measure must express the equivalence of these transactions. However, not only have the institutions of banking not settled upon a system of measurement, they have not even agreed upon what the commodity *is* that they wish to measure; it is this problem that sets the development of markets in ecosystem services apart from more mature markets in easily-quantified commodities such as apples, linen, or even carbon emissions.

Here's the way it's supposed to work according to most bankers, regulators and economic theorists. The value of the banked wetland credit is said to rest in the consumer's desire for an increment of change in a bundle of ecological functions at the bank site, change that happens as the bank site is restored to a presettlement wetland condition by the banker. This "environmental lift" should define the wetland credit. Purchase of this credit is said to compensate for a reduction in the bundle of ecological functions at the site of a wetland impact. Thus, the commodity measures must express the equivalence of ecosystem functions at the site of impact with the ecosystem functions at the bank site: ideally, the customer will buy exactly enough bank credit to cover the lost ecosystem functions at the impact site. Measuring these functions, bank proponents admit, "is one of the most complex issues in mitigation banking." But "once those services are known, they may be translated into a 'currency' which can serve as the medium of trade for a wetland mitigation bank" (ELI 1993: 77).[3]

Measuring Function versus measuring area

Economic theorists have insisted that a standardized technology be used to measure these functions (ELI 1993). Such standardized functional assessments have proliferated since the early 1980s, and are known collectively as "rapid assessment methods" (RAMs). RAMs are usually sets of algorithms, created by wetland scientists, that use easily-measured site characteristics (e.g., plant species diversity or water levels) to make inferences about harder-to-measure "wetland functions" (e.g., habitat provision or peak flow attenuation). Most wetland RAMs use algorithms which translate an empirical observation ("25% canopy cover") into a score ("0.5"), and most produce a series of scores – one for each function. These numeric scores then stand for the wetland. In this way, the commodity sold at a bank is defined in units of incremental ecological function, or environmental lift:

> So we're selling the most perfected environmental lift, under the strictest circumstances, as a method of determining what our product is. Our product is the lift from the pastureland to a thriving, unbelievable wetland with zero exotics on it. That's what our product is.
>
> (Florida bank owner, interview, 2/27/2002)

But in practice the use of RAMs has not resulted in a smoothly-functioning system of valuation. Ideally, to ensure equivalence on both sides of the transaction, RAMs would be used to evaluate functions at both the bank site and the site of impact. At bank sites, however, federal scientists frequently use a combination of RAMs and "best professional judgment" in certifying bank credits for sale (ELI 1993: 88). Thus, quantitative information from RAMs is often used to inform the Corps' certification of credits at a bank, but not in any standardized or reproducible way. Any number of contingencies can introduce unpredictability into what, from a banker's standpoint, should be a very predictable process: "Because if there was ever a discussion" and [the US] Fish and Wildlife [Service] said "Well, you know the Corps says this should be a 'three' but I think it's a 'two' because there really isn't evidence of this [function], it always winds up being a 'two'" (Florida bank owner, interview, 2/27/2002).

The other end of the transaction is no better: the use of RAMs at impact sites is almost nonexistent. Just keeping track of the large number of permit applications and wetland impacts already places a strain on the Regulatory Branch staff of any Corps District. To additionally use a RAM to sort out each wetland impact into its component functional impacts would require an unthinkable commitment of time and expertise for Corps personnel. Therefore, Corps Districts have overwhelmingly chosen to use area, and not function at all, to quantify the impact to a site: functions at impact sites are measured in the imprecise, but easy-to-use, proxy unit of the acre. In the Chicago District of the Corps, for example, under most circumstances, regulations allow that an impact to one acre of natural wetland can be mitigated by the purchase of one certified credit at a mitigation bank.

Markets in segregable ecological functions

This is more or less the current state of affairs in much of the banking industry: functional degradation measured in acres is considered commensurate with functional lift measured using an idiosyncratic mixture of judgment and formalized inference. Bankers, unable to sell their commodity in the units in which they are required to produce it, deeply suspect that the incommensurability built into this arrangement is costing them profit. In short, the "solution" of using acreage as a proxy for function violates market principles, treats impact sites differently from mitigation sites, and satisfies no one. Bankers, using their growing leverage with state and federal policymakers, have begun to lobby to have regulations changed so that all wetland impacts will be assessed "on the same basis on which the bank received its credit evaluation" (Lautin 2001).

This prepares the ground for the neoliberal economist's solution to the problem of commensurability between a wetland impact site and a wetland restoration site, which is to consider each wetland function as a segregable commodity (Kieser 2002; Waters 2002). Thus, if the usage of a RAM reveals that an impact has reduced "hydrologic function" by 3 units, and "duck habitat" by 4 units, the 404 permittee should be able to purchase separate credits, perhaps even at different banks, to mitigate for these impacts. And if, a bank's credit-certification process uses the same RAM, then market actors can have greater faith in the equivalence of exchange values. Bankers can then take advantage of the ecological sophistication of RAMs, by using them to break the mitigation process down into several component transactions. This solution has been enthusiastically embraced by bankers who see the value of their bank multiplying as new, segregated ecosystem services are defined:

> if you can do the same work and "get paid twice", this sure helps. ... It has been shown that a project, which is designed for both functions, can serve the needs of those seeking water retention and wetland restoration. As such, is there any reason not to design the project around an economic model, which produces revenue from both – certainly not.
>
> (Sokolove 2001: n.p.)

In comparison, the Corps' traditional practice of forcing the customer to use an index of value that has no relation to the banker's costs of production is ridiculed as faintly communist.

Conflicting logics

This solution, as some have realized, is a trap: as the logic and language of ecosystem ecology are used more extensively to dictate the definition of ecosystem service commodities, smaller and smaller ecological differences begin to matter, and this creates an entirely new set of problems for commensurability, as some market observers have begun to notice (King 1997; Ruhl and Juge Gregg 2001). For all its faults, the system of ignoring ecosystem information and dictating that "one acre here equals one acre there" provides a unit of measure that offers few geographic barriers to trade. By contrast, incorporating more ecological information may allow different banks the opportunity to reap comparative advantage from their locations and site characteristics; but simultaneously, actual opportunities for trade in increasingly-specialized functions become more restricted. An exasperated EPA official explained this trap of niche markets as a "Pandora's box": the slippery slope that bankers have started down may lead, *ad absurdum*, to a market

> "in habitat for middle-aged great blue herons who don't like shrimp," or something. Obviously, I can't imagine even trying to do that ... you can define a unit so that you're going to have flourishing mitigation banking. You can also define a unit so that, should there ever be one exchanged, it would be environmentally precise. And those are at potentially different extremes.
>
> (interview, 2/28/2002)

Sophisticated RAMs recognize that ecosystem functions are an embedded feature of landscapes, and subject to ecosystem dynamics at many spatial and temporal scales. Therefore, any definition of an ecosystem commodity expressed as a function of ecosystem dynamics will carry with it an implicit argument about the spatial limits to its commensurability. The result is that the further a mitigation bank is away from the site of an impact, the harder it is to argue for commensurability. A vigorous debate has erupted about the appropriate spatial limits to trade in various functions, with some functions (such as duck habitat) seen as "more mobile," and others (such as flood attenuation) seen as "less mobile." Regulators shudder at the prospect of a single bank having, say, seven different geographic service areas for seven different functions. The use of detailed measures of function allows ecologists to argue that it is impossible to move a given function away from its constitutive landscape relations. As an obvious example, it is formally impossible to create the function "floodwater storage for the Kishwaukee River basin" outside the Kishwaukee River basin.

Thus, appealing to the logic of ecosystem science produces a double movement in neoliberal environmental strategy which might be thought of as a version of the tendency for capitalism to oscillate between homogenization and differentiation (Harvey 1990; Smith 1990). On the one hand, banker strategy in appealing for more ecological precision is part of an understandable pursuit of new markets. This suggests that Harvey's description of the geographic dynamism of capital accumulation encompasses the use of ecosystem science in defining and determining differential rent:

> If capitalists become increasingly sensitive to the spatially differentiated qualities of which the world's geography is composed, then it is possible for the peoples and powers that command those spaces to alter them in such a way as to be more rather than less attractive to highly mobile capital.
>
> (Harvey 1990: 295)

On the other hand, the role of ecosystem science in this strategy takes them farther and farther from the kinds of generalizing abstractions that characterize the internal logic of capital, and that are the basic tools of its operation. Acreage-based commensurability is maintained at the cost of ignoring a great deal of ecological information, much as when the infinite variety among individual apples is ignored in pursuit of a unified market in apples. But in using that information to expand opportunities for the circulation of capital, bankers are beginning to use measurements that restrict the actual mechanics of trade, as the uncertain minutiae of ecology become awkwardly consequential for the circulation of capital. One EPA official made a comparison between the prospect of abolishing acres as an abstract measure of "functional lift," and the prospect of abolishing wages as an abstract measure of labor-value:

> Well, it's almost as if we get rid of the dollar bill and we have "my hour of contribution doing X, your hour of contribution doing Y," and et cetera ... we're trying to get at the contribution of function that led to the surrogate of the dollar bill.
>
> (interview, 2/28/2002)

This is a debate over commensurability which could have been lifted straight from Marx's *Capital*. While systems of measuring ecosystem commodities must be functional for capital (they must define a commodity that is alienable, fungible and preferably mobile), they must also be grounded in the naturalized authority of scientific disciplines that are not entirely answerable to the banking industry (Robertson 2006).[4] These requirements seem to guarantee an inconcludable dynamic of contradictory, and perhaps cyclic, impulses in any attempt to constitute markets in ecosystem services.

Conclusion

The vigorous advocacy in policy circles for the development of ecosystem service markets (Daily 2002; Millennium Ecosystem Assessment 2005), and the increasing prominence of market-led environmental policies (USEPA 2003; USDA 2005), demonstrate the abundant energy in the public, private and nonprofit sectors that drives neoliberal policy. Policymakers, scientists, and entrepreneurs seem to be taking Costanza's imperative seriously. The story of banking shows both the success and limits of the neoliberal strategy of requiring scientists to provide appropriate and useful measurement technologies, and may point to difficulties that will be experienced by future neoliberal environmentalisms. As the problem of measurement shows, the attempt to push ecological knowledge towards spawning further rounds of accumulation may disrupt the very mechanics of accumulation. In observing this, I have found that it does not pay to treat the particulars of ecological science with any less attention than the particulars of labor relations or industrial processes as they factor into state strategy or capital circulation. And yet this is not an impatient call for "sound science": it is simply to recognize that some ecological knowledges "work" for capital and that some do not (Braun 2000), and that this intransigence (or "uncooperativeness" as Bakker (2005) has very usefully characterized it) may be an unexpected clue to neoliberalism's vulnerabilities. Neoliberal strategy appears much less potent when its use of ecological science is revealed as one more unstable, crisis-ridden moment. Thus, one's approach to the concept of "nature" has everything to do with how one will understand crisis and resistance in neoliberal environmentalism. Rather than accepting the neoliberal conceit of nature as an external surface, economic and political geographers must study the way ecological science achieves the effect of materiality in describing a "territory with qualities," because this achievement is essential in allowing economic and political strategies to effectively play out across that territory (Braun 2000). Alternatively, if we insist on understanding nature as either fully determined by the cultural and economic forces of capitalism, or as having a robustly material ontology, or as some arbitrarily segregated combination of the two, then we have little choice but to commit to some form of environmental determinism or to be altogether silent on the subject.

Acknowledgments

A longer version of this chapter was Robertson, M. (2004) "The Neoliberalization of Ecosystem Services: Wetland Mitigation Banking and Problems in Environmental Governance," *Geoforum* 35 (3): 361–73.

Notes

1 The Bush and Clinton Administrations were in perfect agreement on the subject of banking, which was promoted in Clinton's wetland policy as well (WHOEP 1993).
2 Pseudonyms have been used by request in some instances cited in this chapter.
3 I am leaving aside the more basic (and realist) notion of whether or not the restored sites are actually providing the ecosystem services that policymakers assume they do, given the relative youth of the science of ecosystem restoration. In the context of the current discussion, it is important to realize that the hesitancies and caveats of restoration ecologists do not articulate well with the capital's need for naturalized and authoritative scientific knowledge.

4 One might well ask to whom scientists *are* answerable: how physical do we want to get in specifying the "ground" in which science's naturalized authority is grounded? It is of course extremely tempting to take the critical realist position that normal science has some mediated, but falsifiable, access to real nature, which is seen to explain normal science's stability and precision. I have argued (Robertson 2004) against this position; scientists' ability to resist capital logics is grounded in the stability and coherence of a system of knowledge production, not in their access to real nature. The notion that letting go of the security of the realist "teddy bear" is to embrace a rudderless social constructivism is one of the most pernicious false dichotomies going in social science (Latour 1993).

References

Bakker, K. (2005) "Neoliberalizing Nature? Market Environmentalism in Water Supply in England and Wales," *Annals of the Association of American Geographers* 95(3): 542–65.

Braun, B. (2000) "Producing Vertical Territory: Geology and Governmentality in Late Victorian Canada," *Ecumene* 7(1): 7–46.

Castree, N. (2002) "False antitheses: Marxism, nature and actor-networks," *Antipode* 34 (1): 119–48.

Coase, R. (1960) "The Problem of Social Cost," *Journal of Law and Economics* 3: 1–44.

Costanza, R., d'Arge, R., de Groot, R., Farber, S., Grasso, M., Hannon, B., Limburg, K., Naeem, S., O'Neill, R. V., Paruelo, J., Raskin, R. G., Sutton, P., and van den Belt, M. (1997) "The Value of the World's Ecosystem Services and Natural Capital," *Science* 387(6630): 253–60.

Daily, G. C. (1997) "Introduction: What Are Ecosystem Services?" In G. C. Daily (ed.) *Nature's Services: Societal Dependence on Natural Ecosystems*, Washington, DC: Island Press, pp. 1–10.

Daily, G. C. and Ellison, K. (2002) *The New Economy of Nature*, Washington, DC: Island Press.

Environmental Law Institute (ELI) (1993) *Wetland Mitigation Banking*, Washington, DC: Environmental Law Institute.

Goldman, M. (1998) "Inventing the Commons: Theories and Practices of the Commons' Professional," in M. Goldman (ed.) *Privatizing Nature: Political Struggles for the Global Commons*, New Brunswick, NJ: Rutgers University Press, pp. 20–53.

Harvey, D. (1990) *The Condition of Postmodernity: An Enquiry into the Origins of Cultural Change*, Malden, MA: Blackwell.

Heal, G. (2000) *Nature and the Marketplace: Capturing the Value of Ecosystem Services*, Washington, DC: Island Press.

Henderson, G. (1998) "Nature and Fictitious Capital: The Historical Geography of an Agrarian Question," *Antipode* 30(2): 73–118.

Hockenstein, J. B., Stavins, R. N. and Whitehead, B. W. (1997) "Crafting the Next Generation of Market-Based Environmental Tools," *Environment* 39(4): 13–20, 30–3.

Kieser, M. (2002) "Developing Markets to Manage Ecosystems," paper presented at the 5th National Mitigation Banking Conference, Washington, DC, March 1, Terrene Institute.

King, D. M. (1997) "The Fungibility of Wetlands," *National Wetland Newsletter* 19(5): 10–13.

Latour, B. (1993) *We Have Never Been Modern*, Cambridge, MA: Harvard University Press.

Lautin, L. (2001) "Roles and Responsibilities of Mitigation Banking," paper presented at the 4th National Mitigation Banking Conference, Ft. Lauderdale, FL, April 19, Terrene Institute.

McEvoy, A. F. (1986) *The Fisherman's Problem: Ecology and Law in the California Fisheries, 1850–1980*, Cambridge: Cambridge University Press.

Millennium Ecosystem Assessment (2005) *Ecosystems and Human Well-Being: Synthesis*, Washington, DC: Island Press.

Robertson, M. M. (2004) "The Neoliberalization of Ecosystem Services: Wetland Mitigation Banking and Problems in Environmental Governance," *Geoforum* 35(3): 361–73.

—— (2006) "The Nature That Capital Can See: Science, State and Market in the Commodification of Ecosystem Services," *Environment and Planning D: Society and Space* 24(3):

Ruhl, J. B. and Juge Gregg R. (2001) "Integrating Ecosystem Services into Environmental Law: A Case Study of Wetlands Mitigation Banking," *Stanford Environmental Law Journal* 20: 365–92.

Smith, N. (1990) *Uneven Development: Nature, Capital and the Production of Space*, Oxford: Blackwell.

Sokolove, R. D. (2001) "Multiple Uses for Wetland Mitigation Banks," paper presented at the 4th National Mitigation Banking Conference, Ft. Lauderdale, FL, April 20, Terrene Institute.

St. Martin, K. (2001) "Making Space for Community Resource Management in Fisheries," *Annals of the Association of American Geographers* 91(1): 122–42.

Taylor, P. J. (1997) "How Do We Know We Have Global Environmental Problems? Undifferentiated Science-Politics and Its Potential Reconstruction," in P. J. Taylor, S. E. Halfon and P. N. Edwards (eds.) *Changing Life: Genomes, Ecologies, Bodies, Commodities*, Minneapolis, MN: University of Minnesota Press, pp. 149–74.

Tietenberg, T. H. (1985) *Emissions Trading: An Exercise in Reforming Pollution Policy*, Washington, DC: Resources for the Future, Inc.

US Department of Agriculture (USDA) (2005) *Secretary's Memorandum: USDA Roles in Market-based Environmental Stewardship*, Washington, DC: USDA.

US Environmental Protection Agency (USEPA) (2003) "Water Quality Trading Policy," available at: www.epa.gov/owow/watershed/trading/finalpolicy2003.pdf.

Waters, S. (2002) "Issues in the Economic Appraisal of Mitigation," paper presented at the 5th National Mitigation Banking Conference, Washington, DC, February 28, Terrene Institute.

White House Office on Environmental Policy (WHOEP) (1993) *Protecting America's Wetlands: A Fair, Flexible and Effective Approach*, Washington, DC: The White House.

10 Weak or strong multifunctionality?

Agri-environmental resistance to neoliberal trade policies

Gail Hollander

The Uruguay Round Agricultural Agreement of 1993 has been recognized as one of the "defining moments" (Potter and Ervin 1999: 53) of twentieth-century agricultural policy reform. Emphasizing the link between domestic agricultural policies and international trade, it committed signatories to reducing domestic subsidies, except for support provisions that are categorized as non-trade-distorting. These, termed "Green Box" policies, were "hotly debated in the negotiations" (Josling *et al.* 1996: 206). Indeed, looking inside the Green Box reveals room for interpretation regarding the impact of the Uruguay Round on the liberalization of agricultural trade.

One of the new approaches to agricultural regulation being debated for inclusion in the Green Box is "multifunctionality," originating in the European Union (EU). While this chapter will explore various meanings of this contested term, in broadest terms it refers to the idea that agricultural landscapes may serve multiple functions, preserving biodiversity, culture, and livelihoods and, of course, producing agro-commodities. The term also signals support for policies designed to protect these multiple functions. Emerging in the context of agricultural trade liberalization and reform of the Common Agricultural Policy (CAP), multifunctionality is promoted as a way to address social and ecological concerns such as farm abandonment and biodiversity loss through domestic agricultural policies that conform to the GATT/World Trade Organization (WTO). Multifunctionality is a response to what are seen to be the negative aspects of trade liberalization and could thus be described as a form of resistance that has been formulated at scales ranging from supranational to national to very local.

The analyses presented here are informed by the literature on the globalization of the agro-food system. Issues of global governance became increasingly salient in the 1990s, a period of "unprecedented deregulation of agriculture (a shift from aid to trade), the hegemony (the so-called 'new realism') of export-oriented neoliberal development strategies, and a recognition that globalisation ... of the world agro-food economy was proceeding apace" (Watts and Goodman 1997: 1). This chapter considers multifunctionality as a political strategy and policy instrument currently being deployed to negotiate the particularities of place-based production systems within the realm of neoliberal agro-food politics.

I first consider multifunctionality as one response to neoliberal pressures for agricultural reform. Second, I explore the possibility for multifunctionality to serve as a vehicle for resistance to GATT/WTO in other world regions. I do this through a study of arguably the most maligned agricultural zone in the world, the sugar-producing region of south Florida. The geographic focus is the Everglades Agricultural Area (EAA) that is home to the Florida "Sugar Bowl" (see Figure 10.1). Viewing the landscape of Florida sugar through the lens of multifunctionality, I examine the possibilities and limitations in the geographic transferability of multifunctionality.

I begin by examining the way that EU agricultural advocates, in their opposition to neoliberal trade policies, have used ideas of landscape, livelihood and agroecology, encompassed by the term "multifunctionality," in defense of domestic agricultural supports, and then identify what I term the "weak" and "strong" versions. In the following section, I turn to the case

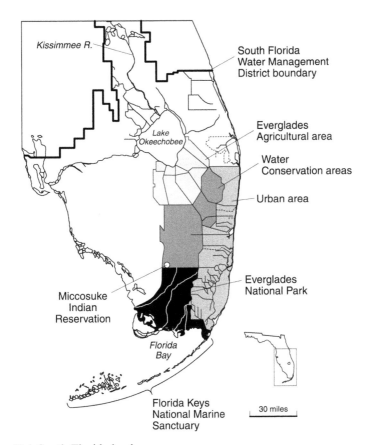

Figure 10.1 South Florida land use.

study of sugarcane production in south Florida. I conclude with a discussion of the transferability of multifunctionality to this case and the wider geographic implications of placing Florida sugar in the Green Box.

The emerging concept of multifunctionality in the EU

The Uruguay Round Agricultural Agreement brought agriculture more firmly under the purview of the GATT and committed signatories to the rule-based, market liberalization of the WTO, which convened an Agricultural Committee for trade negotiations beginning in March 2000. In response to this push for trade liberalization, as well as to the reintegration of Eastern European agriculture, EU policymakers articulated a multifunctional "European Model of Agriculture" in a series of policy documents published during the late 1990s. The EU Agricultural Commissioner defined "multifunctionality" as the link "between sustainable agriculture, food safety, territorial balance, maintaining the landscape and the environment and what is particularly important for developing countries, food security" (quoted in Buller 2001: 4). Thus, "multifunctionality" entered the lexicon of globalization at the close of the century as part of the conceptual apparatus and the discursive strategies deployed to debate and negotiate neoliberal agricultural trade policies in domestic and international fora.

Because the concept emerged in defense of the perceived particularities of European rurality, it has been characterized as "a model that reflects the specific history, cultures and choices of European society" (Givord 2000). In contrast, Potter and Burney are less geographically-specific: "The central assumption of this model is that agriculture is multifunctional, producing not only food but also sustaining rural landscapes, protecting biodiversity, generating employment and contributing to the viability of rural areas" (Potter and Burney 2002: 35).

In the parlance of WTO negotiations, "multifunctionality" provides the philosophical underpinnings to argue for the expansion of the Green Box. Multifunctionality provides a strategic opening in which to recognize the landscape functions of agriculture and rural settlement, so that the resultant social and ecological complexity can be defined as public goods and maintained through state policies. It represents a shift in emphasis from the negative to the positive environmental externalities of agricultural production to argue for recognition of the social and/or environmental goods that are "jointly produced" along with agricultural products.

The response at the international level has varied according to the interests and alignments of various states. Japan, South Korea, Norway and Switzerland, with the EU have formed the "Friends of Multifunctionality" to emphasize "non-trade" aspects of agricultural production in multilateral negotiations. In contrast, multifunctionality has provoked a critical response from the Cairns Group,[1] which regards it as "a smokescreen for the continuation of protectionist agricultural policies" (Potter and Burney 2002: 36).

While some characterize the US response as skeptical (Givord 2000; Freshwater 2001), others have termed it "equivocal" (Potter and Burney 2002),[2] a reflection of the contradictory position it holds, officially advocating free trade while maintaining protectionist agricultural policies. However, the US Department of Agriculture report (Bohman *et al.* 1999) on multifunctionality omits reference to cultural and ecological diversity, reducing agri-environmental goods to "scenic vistas."

There has been a flurry of interest in recent years in the idea of multifunctionality, as evidenced by the burgeoning list of conferences, reports, publications and websites devoted to the topic. It is controversial and contentious, affording an element of protection – either protectionist trade policy or protective of valued landscapes, communities, and ecological services – depending on one's viewpoint. Invoked in defense of low intensity farming systems with high nature conservation values associated with *traditional* patterns of human interference, it is being articulated as an anti-development or alternative development discourse emanating from the "North." In analyzing the various conference documents, alliances for and against multifunctionality, and government and NGO position statements, I have identified two distinct versions of the concept, which I label "weak" and "strong" multifunctionality. Weak multifunctionality defends a limited set of national interests in the agricultural sector. Strong versions challenge the current structure and logic of trade liberalization as regulated by GATT agreements and the WTO and view multifunctionality as a path to radical reform. The case of the FAO/Netherlands September 1999 "Cultivating our Futures" Conference and the debates that followed will serve to illustrate.

The FAO/Netherlands conference was organized as a pre-conference to implement the goals of Agenda 21 for sustainable agriculture and the World Food Summit. Because multifunctionality had developed into a highly contentious term in ongoing trade negotiations, the organizers proposed "Multifunctional Character of Agriculture and Land" (which became "MFCAL") as an alternative. However, representatives from Cairns Group countries rejected this distinction, protesting that MFCAL and multifunctionality were identical, protectionist, and "seriously flawed as a concept" (Doran *et al.* 1999: 3).

Following the conference and in preparation for Earth Summit 2002, the French non-governmental organization SOLAGRAL (Solidarité Agricole et Alimentaire) posted a caucus position paper asking "Is it worth defending the concept of multifunctionality of agriculture?" (Jadot 2000). The author rejected the weak version, denying European exceptionalism: "There is no 'European farm model', there is an export-oriented model that needs to become a multifunctional one" and challenged the strategy of tinkering with the "green box." He criticized the narrow alliance comprising the "Friends," concluding that a defensible multifunctional grouping would include developing countries, "first and foremost those that face food security problems."

Agro-industry in the Everglades

In contrast to Europe's "traditional agricultural landscapes" shaped over centuries (Vos and Meekes 1999), the landscape of sugarcane cultivation in south Florida represents a rupture with the historic Everglades landscape it overlays. Transformation occurred in decades – from the inception of commercial sugar production in the 1920s to the completion in 1972 of the drainage system mandated in the Central and South Florida Flood Control Act of 1948. The Act designated the Everglades Agricultural Area (EAA), 700,000 acres of muck soils drained for commercial agriculture and devoid of features that the USDA might consider a "scenic vista." Thus neither the "European model" of a palimpsest of agricultural history nor the US model of a scenic vista is found in the south Florida sugar bowl.

Venturing into the EAA is not unlike entering the "modern giant factory building" envisioned by early promoters of Everglades agriculture. The landscape is reflective of the effort to maximize mass production efficiency. Roads and rails, oriented to the mills, bear trucks and trains laden with cane. The logic of sugar production shapes the geography of human settlement. Throughout the EAA are corporate-owned plantation villages, housing the plantation workforce, equipment and maintenance buildings for the surrounding landscape. The EAA is thus a vast agro-industrial territorial production complex, including 430,000 acres planted in sugarcane, six mills, and two sugar refineries. Two corporations, Flo-Sun Incorporated and United States Sugar Corporation (USSC), account for 190,000 and 160,000 acres of sugarcane respectively, and five of the six mills.[3]

Despite its factory-like appearance, issues of cultural landscape, livelihoods, and agroecology *are* important in the EAA, though "Big Sugar" dominates the popular imagination. However, Big Sugar – which refers to the large vertically integrated grower-processors – has coexisted with "Little Sugar" since the 1940s, when farmers in the area were encouraged to diversify. Together, Florida's large and smaller growers produce 52 percent of the US supply of domestically produced cane sugar.

The relationship between the sugar corporations and independent growers in the region is necessary and symbiotic. Interviews with family farmers demonstrate this interdependency.[4] Farmers need to have a contract to assure their cane will be milled: "Before I planted my first stick of cane I had to go to Sugar (USSC), 'Will you grind?' You have to get a home for it. It isn't like oranges, where you can tote 'em and sell 'em" (Kirk 1995).[5] This farmer represented a multigenerational farming operation dating to the 1940s, with 960 acres in cattle, 1280 acres in cane, and 340 acres in citrus. They diversified from cattle in 1984, using sugar to stabilize their income, a common choice in the EAA. Another farmer, whose family had farmed in the area since 1915, explained why he had helped found the Sugar Cane Growers Cooperative in 1960: "I was in the vegetable business and my family before me. It's hit and miss. There's no stability in the produce business. Sugar is more stable because of our sugar policy. I wanted to add some stability to my operations, so when this opportunity came along, we took it" (Schmidt 1995).

Some families have been raising sugar since the 1940s, when federal production controls were lifted and farmers were encouraged by USSC to plant cane. A member of one such family explained how two households and two full-time employees were supported on 650 acres of sugarcane: "My aunt paid off the land a long time ago. She owns the land and she makes the big decisions" (Walker 1995). He noted that when a freeze killed the cane in the fields, USSC stopped its own harvesting to save small farmers' crops. Another farmer, the primary operator of a 500-acre farm, explained how, when the family dairy business failed in the 1980s, she learned to cultivate sugarcane: "The sugar company helped us a lot when we first got started. I was scared slap to death. I knew nothing whatsoever. Thank god for an old guy that worked for US Sugar, he taught me how to raise cane" (Bentley 1995).

Ethnographic fieldwork informed my evaluation of multifunctionality in the EAA. Many elements of multifunctionality – such as cultural landscape, cultural heritage, generating employment and contributing to the viability of rural areas – are evident in the way that people express their ideas about the landscape, environment, and differences between rural and urban Florida. Living among and listening to EAA residents, one can identify their strong sense of identity rooted in place. Many of them are second and third generation Floridians (unusual in the Sunshine State), descendants of immigrants from the US South or the Caribbean. They view themselves as having environmental knowledge and a set of values different from those of the Florida urban masses, more recently arrived from the Midwest and Northeast US or from Latin America. Conversely, coastal residents who are unfamiliar with the EAA view it as the terrain of Big Sugar, which they single out as an obstacle to Everglades restoration. The EAA community was galvanized in 1994 when the Science Sub-Group of the Everglades Task Force suggested as a possible ecosystem restoration measure, "the return of wetland function to all former wetlands in current agricultural uses, including the entire Everglades Agricultural Area." Over the next few years, environmental groups attempted to pass a referendum to place a special tax on sugar produced in Florida. At that time, environmental groups thought phosphorous run-off from the EAA was responsible for the eutrophication of Florida Bay and deterioration of offshore reefs. The feeling of injustice from being blamed for environmental problems was revealed as a division between urban and rural, or, more specifically, coastal and central Florida. "Why is it that we're the only ones [causing environmental problems] when those automobiles and all that pollution over on the coasts, the septic tanks – they accused us of ruining Florida Bay!" (Stoll 1995).

A USSC plantation foreman noted the irony of suburban dwellers condemning agriculture as ecologically destructive, revealing knowledge of the relative ecological benefits of land uses and the historical relation between recent suburban settlement compared to earlier agricultural development:

One person, he wanted to take the Glades back to that original environment, which of course is impossible to do, and I asked this person where they lived and they lived in Wellington. I said, "Well, you know, twenty years ago all of Wellington was under water, it was a swampland. I think that oughta be returned to its natural state." "Well," they said, "that's not the same!" I don't know, I didn't see the difference.

(Richards 1995)

Often lost in the debates over the south Florida environment is the ecological role of agriculture versus available alternatives. In the EU, advocates of multifunctionality emphasize the ecological services of existing land management regimes. In the US environmentalists have tended to pose agriculture as the main ecological problem in south Florida. While not familiar with multifunctionality, some of the EAA farmers noted an important policy difference between Europe and the US:

The environmental groups in England particularly, but I guess in most of Europe, have done a better job of preserving wildlife and things cause they went at it where they made it to the benefit of the farmer to preserve it. Over there you get tax breaks and advantages if you preserve things. If I go out here and find something, it's to my detriment that I've got a protected species on my property.

(Kirk 1995)

Many farmers questioned the wisdom of state acquisition and retirement of agricultural land. Their comments on this topic were informed by their knowledge of the landscape, which led them to conclude that idled land would be susceptible to invasion by exotic species that would create cover inhospitable to wildlife. One farmer noted how elsewhere in Florida sandhill cranes had moved into agricultural areas out of former grazing lands that were now state-managed preserves, "because they can't make it in the brush, you know" (Kirk 1995). Another farmer identified in his comments two species that the state of Florida has labeled "invasive exotics":

I don't think they can take care of what they own now. Melaleuca trees and Brazilian pepper trees are growing on state land now, where if someone still farmed it, or was grazing it at least, it would be in good enough shape to not have exotics growing on it.

(Walker 1995)

Florida sugar as a multifunctional agro-environment

Central to any claim for multifunctionality is that a particular agricultural system provides ecological services "jointly produced" with agro-commodities.

This would seem an unlikely assertion to be made on behalf of the sugar-cane agro-industry, which environmental organizations, such as "Friends of the Everglades," suggest should "move" from the EAA to further Everglades ecological restoration. To consider this aspect of multifunctionality requires some background regarding the ecological issues of the Everglades and the EAA in particular.

The original Everglades ecosystem was hydrologically contiguous, 7500 square miles stretching from Lake Okeechobee to Florida Bay, comprised of a mosaic of habitats. Due to seasonal rainfall, this extensive wetland was characterized by a pulsed sheet flow coursing over limestone bedrock. Most importantly, it was an oligotrophic system, limited by phosphorous. Today, about half the original Everglades remains and concerns center on altered hydrology, including loss of water storage, overland flow and spatial extent; loss of connectivity among fragmented wetlands; changes in hydroperiod and fire regime; changes in water quantity and quality; and invasion of exotic species.

A significant environmental problem stems from the fact that the drained soils of the EAA oxydize, decompose and subside, at an average rate of one inch per year in the EAA. Oxydation is directly proportional to the depth of the water table, so that a deeper water table means a greater rate of subsidence. The problems caused by subsidence are that the loss of soils threatens the future of agriculture and that as soils decompose, nitrogen and phosphorous are mineralized and released into the environment.

The problem of subsidence is thus related to one of the key issues defining the relation between agriculture in the EAA and the remaining Everglades: the quality of water discharged from the EAA, specifically the level of phosphorous. Phosphorous-laden run-off from the EAA is blamed for eutrophication and subsequent shift in species composition in adjacent marshes (McCormick *et al.* 2002; Sklar *et al.* 2002). The 1992 Everglades Best Management Practices (BMPs) Program mandated a 25 percent reduction in phosphorous in run-off from the EAA. In experiments to develop BMPs, researchers found that phosphorous concentrations in drainage water were lower from sugarcane versus fallow drained plots because more phosphorous left the fields in biomass than was applied (Izuno *et al.* 1995).

A USDA agronomist expressed grave reservations regarding large-scale restoration based on retiring agricultural land. To him, the key to agricultural sustainability and ecological restoration lay in addressing the problem of soil subsidence:

> If we can restore the natural hydrology and still have a productive sugarcane crop, then, by concentrating on controlling soil subsidence and having a productive agriculture, we're meeting all the bum raps against the EAA. Soil oxidation is actually the major source of phosphorous. So if we can control that we're coming up with *the* major long-term solution for phosphorous control.

(Baker 1996)

Since the key to reducing or eliminating subsidence is to raise the water table, the implication for sugarcane genetics is to breed for flood tolerance. Recent research on its root structure had revealed sugarcane to be a flood-tolerant species, which had implications for Everglades restoration. Because the historical water regime included dry periods that correspond to the periods in which key farming operations must be undertaken, the agricultural system could mimic historic hydrological systems:

> [W]e felt that in order to control subsidence the EAA would have to be flooded all the time, twelve months out of the year. But what I learned from the hydrologists and the ecologists was in fact, it would not have been flooded during the dry season, which just so happens to correspond to the sugarcane planting and harvesting season.
>
> (Baker 1996)

In assessing the ecological role of sugar in the landscape, the point here is quite adamantly *not* that the Everglades should have been drained for sugar. The question now is, how does it compare to alternative land uses that are possible in this irreversibly transformed landscape? Sugarcane offers certain ecological benefits such as the economic incentive and wherewithal to manage exotic vegetation, the possibility of mimicking water regimes associated with the historic Everglades, and the ability to maintain a landscape mosaic. In the opinion of a wildlife biologist who specializes in the study of alligators and crocodiles as indicator species, "You get a lot more – you get a lot better combination of ecological benefits, or at least the potential for ecological benefits and economic productivity, with agriculture" (Jones 1996).

Discussion

This assessment of livelihoods, landscape, and agroecology in the Florida EAA raises a number of questions regarding the concept of multifunctionality, including its meaning, geographic transferability, coherence, and implications for agricultural trade liberalization. Though the debate surrounding the political-economic and geographical limits of multifunctionality will continue, we can observe that the weak version will be increasingly harder to defend. The argument for environmental goods and services has been made on the basis of European experience, that EU farming has "been compatible with the conservation of biodiversity and other environmental benefits" and "may also have actively molded their very character through a process of joint production of food and environmental goods" (Potter and Burney 2002: 39). This informs the philosophical position that there is a distinctive "European model of agriculture." These claims of the case for European agriculture as distinct from the rest of the world do not withstand scrutiny. Europe is just one example, and not even the most ancient, of many other regions in Asia, Africa, the Middle East, and Mesoamerica, where land has been farmed for millennia. As for European agriculture providing environmental services as well as agricultural goods, the case of the Florida EAA demonstrates that this is not particularly unusual or distinctive.

To return to the case of Florida, what insights does the Florida example have to offer in terms of multifunctionality? At first glance, Florida sugar seems an unlikely example of a multifunctional agricultural landscape. Agriculture in the EAA can neither be characterized as "more traditional" (Knickel and Renting 2000: 526) nor as "marginal" (Potter and Goodwin 1998: 291). Yet EU concerns, such as "sustaining rural landscapes, protecting biodiversity, generating employment and contributing to the viability of rural areas" (Potter and Burney 2002: 35) find parallels in this region. The EAA is a distinctive rural region, located between two of the fastest growing and urbanizing coastal regions in the US. Residents of the EAA feel their agricultural livelihoods are endangered by demands for water storage and threatened by efforts to link trade liberalization with Everglades restoration. A closer look at Big Sugar reveals a differentiated picture, including several large corporations, thousands of agro-industrial workers, and a range of small and medium-sized diversified farming operations, for which sugar provides a relatively stable income stream. As Snyder and Davidson note, "If the industry were to collapse, ... small landowners, including families that have farmed the area for years and even generations, also would likely be bankrupted" (1994: 111).

As we turn to the agri-environmental question, it has to be asked in reference to the landscape of the present and with regard to the future. The sugar industry has greatly transformed the pre-human and early human landscape, but by definition, so has every other agricultural region. The question is not whether it is ecologically and economically sensible to drain the Everglades for sugar but what is the function of sugarcane now? In formulating their response to agri-environmental issues farmers, scientists, and industry representatives are beginning to develop a *local* knowledge base regarding the regional landscape dynamics. With regard to the future, Glaz suggests:

> As the EAA evolves to a zero-subsidence agriculture, it would also be evolving to conditions more similar to its natural predrainage conditions. Hydrologists may be able to take advantage of these new conditions to reestablish the natural hydrological links of the EAA.
>
> (1995: 611)

The concept of multifunctionality emerged in the context of increasing pressure from the WTO perceived as threatening to rural land-based economies and their associated landscapes. The weak version does not challenge the overall course of globalization and trade liberalization, but seeks rather to create a space for European agriculture by stressing its non-trade characteristics. The weak version appears to be transferable to the Florida case when we think in terms of the domestic politics.

Does the case of Florida fit the "strong" version of multifunctionality? The strong version could provide the conceptual framework for countries and regions to resist efforts to implement free trade initiatives. However, several problems arise when we apply the "strong" version to this case. If the "strong" version includes biodiversity, cultural preservation, food security, and sustainable development in an effort to address the concerns of both North and South, then we see that not all aspects would apply to Florida. The difficulty in applying the "strong" version of multifunctionality to the case of Florida sugar points to two conceptual difficulties. First, even its supporters admit that it is mutable. By attempting to pull into a common framework such disparate issues as food security and the preservation of stone fences, the term loses coherence. Yet to the extent that it does convey that there are special qualities inherent in land-based food and fiber production systems and common concerns regarding rural marginalization, we encounter the second difficulty of the term, which is that it has become the focus of struggle over agricultural exceptionalism. Multifunctionality has become the synecdoche of international agricultural politics to the extent that the acceptance of the term is clearly guided by how a country sees its interests in agricultural trade liberalization. The question of whether the US would use the concept of multifunctionality in international agricultural trade negotiations in general or with respect to sugar remains to be seen. Though some suggest that US farm policy is moving in the direction of multifunctionality (see McCarthy 2005, for discussion), there is not yet evidence that this is done in the comprehensive way necessary to provide moral standing in international trade negotiations. That is, it is not clear that US trade negotiators are willing to recognize the legitimacy of multifunctionality. Meanwhile, the goal of more ecologically sustainable production may well be undone by the increasing emphasis on agrocommodities such as corn, soy and sugar as a source of biofuels. Moreover, the collapse of the Doha Round of global trade talks, attributed to the power of US and EU farmers to forestall meaningful reform of their respective domestic agricultural programs, suggests that the impetus to pursue strong multifunctionality is lacking.

Acknowledgments

A longer version of this chapter was Hollander, G. (2004) "Agricultural Trade Liberalization, Multifunctionality, and Sugar in the South Florida Landscape," *Geoforum* 35(3): 299–312.

Notes

1 The Cairns Group formed in 1986 to advocate agricultural trade liberalization, comprising 17 agricultural exporting countries (Argentina, Australia, Bolivia,

Brazil, Canada, Chile, Colombia, Costa Rica, Guatemala, Indonesia, Malaysia, New Zealand, Paraguay, the Philippines, South Africa, Thailand and Uruguay) that together account for one-third of the world's agricultural exports.

2 The US negotiating proposals that were tabled at the June 2000 WTO Agricultural Committee Meeting included additional domestic support criteria – such as environmental and natural resource protection – that were welcomed by the EU as resembling multifunctionality and expanding the green box.

3 The sixth mill is owned by the Sugar Cane Growers Cooperative of Florida, which was chartered by area vegetable growers in 1960.

4 Ethnographic fieldwork was conducted at different periods from 1995 through 1996, and included extensive interviews with farmers, workers, company representatives and scientists.

5 Names have been changed in order to protect anonymity.

References

Bohman, M., Cooper, J., Mullarkey, D., Normile, M., Skully, D., Vogel, S. and Young, E. (1999) *The Use and Abuse of Multifunctionality*, Washington, DC: Economic Research Service/USDA.

Brouwer, F. and Straaten, J. van der (2002) "Agriculture and Nature in Conflict?" In F. Brouwer and J. van der Straaten (eds.) *Nature and Agriculture in the European Union*, Cheltenham: Edward Elgar.

Buller, H. (2001) "Is This the European Model?" In H. Buller and K. Hoggart (eds.) *Agricultural Transformation, Food and Environment: Perspectives on European Rural Policy and Planning*, vol. 1, Aldershot: Ashgate Publishing.

Doran, P., Schmidt, K. and Spence, C. (1999) "Summary of the Conference on the Multifunctional Character of Agriculture and Land," *Sustainable Developments* 32(5): 1–10.

Freshwater, D. (2001) "A U.S. Perspective on Multifunctionality," paper prepared for the World Bank Workshop on Multifunctionality, Washington, DC, available at: www.worldbank.org/wbi/B-SPAN/sub-multifunction1.htm.

Givord, D. (2000/2001) "Defending the European Rural and Agricultural Model at the WTO," *LEADER Magazine* 25(Winter).

Glaz, B. (1995) "Research Seeking Agricultural and Ecological Benefits in the Everglades," *Journal of Soil and Water Conservation* 50(6): 609–12.

Izuno, F. T., Bottcher, A. B., Coale, F. J., Sanchez, C. A. and Jones, D. B. (1995) "Agricultural BMPs for Phosphorous Reduction in South Florida," *Transactions of the American Society of Agricultural Engineers* 38(3): 735–44.

Jadot, Y. (2000) "Is It Worth Defending the Concept of Multifunctionality of Agriculture?" Contribution of SOLAGRAL to the Intersession of Conference on Sustainable Development 8 available at: www.csdngo.igc.org/agriculture/agr_solagral.htm

Josling, T., Tangermann, S. and Warley, T. (1996) *Agriculture in the GATT*, New York: St. Martin's Press.

Knickel, K. and Renting, H. (2000) "Methodological and Conceptual Issues in the Study of Multifunctionality and Rural Development", *Sociologia Ruralis* 40(4): 512–28.

Lowe, P., Buller, H. and Ward, N. (2002) "Setting the Next Agenda? British and French Approaches to the Second Pillar of the Common Agricultural Policy," *Journal of Rural Studies* 18: 1–17.

McCarthy, J. (2005) "Rural Geography: Multifunctional Rural Geographies – Reactionary or Radical?" *Progress in Human Geography* 29(6): 773–82.

McCormick, P. V., Newman, S., Miao, S., Gawlik, D. E., Marley, D., Reddy, K. R. and Fontaine, T. (2002) "Effects of Anthropogenic Phosphorous Inputs on the Everglades," in J. W. Porter and K. G. Porter (eds.) *The Everglades, Florida Bay, and Coral Reefs of the Florida Keys: An Ecosystem Handbook*, Boca Raton, FL: CRC Press.

Noe, G. B., Childers, D. L. and Jones, R. D. (2001) "Phosphorous Biogeochemistry and the Impact of Phosphorous Enrichment: Why Is the Everglades So Unique?" *Ecosystems* 4: 603–24.

Potter, C. and Burney, J. (2002) "Agricultural Multifunctionality in the WTO: Legitimate Non-trade Concern or Disguised Protectionism?" *Journal of Rural Studies* 18: 35–47.

Potter, C. and Ervin, D. (1999) "Freedom to Farm: Agricultural Policy Liberalisation in the US and EU," in M. Redclift, J. Lekakis, and G. Zanias (eds.) *Agriculture and World Trade Liberalisation: Socio-environmental Perspectives on the Common Agricultural Policy*, Wallingford: CABI Publishing.

Potter, C. and Goodwin, P. (1998) "Agricultural Liberalization in the European Union: An Analysis of the Implications for Nature Conservation," *Journal of Rural Studies* 14(3): 287–98.

Rice, R. W., Izuno, F. T. and Garcia, R. M. (2002) "Phosphorous Load Reductions under Best Management Practices for Sugarcane Cropping Systems in the Everglades Agricultural Area," *Agricultural Water Management* 56: 17–39.

Sklar, F., McVoy, C., Van Zee, R., Gawlik, D. E., Tarboton, K., Rudnick, D., Miao, S. and Armentano, T. (2002) "The Effects of Altered Hydrology on the Ecology of the Everglades," in J. W. Porter and K. G. Porter (eds.) *The Everglades, Florida Bay, and Coral Reefs of the Florida Keys: An Ecosystem Handbook*, Boca Raton, FL: CRC Press, pp. 39–82.

Snyder, G. H. and Davidson, J. M. (1994) "Everglades Agriculture: Past, Present, and Future," in S. M. Davis and J. Ogden (eds) *Everglades: The Ecosystem and its Restoration*, Delray Beach, FL: St. Lucie Press.

Vos, W. and Meekes, H. (1999) "Trends in European Cultural Landscape Development: Perspectives for a Sustainable Future," *Landscape and Urban Planning* 46: 3–14.

Watts, M. and Goodman, D. (1997) "Agrarian Questions: Global Appetite, Local Metabolism: Nature, Culture, and Industry in Fin-de-Siècle Agro-food Systems," in D. Goodman and M. Watts (eds.) *Globalizing Food: Agrarian Questions and Global Restructuring*, London and New York: Routledge.

11 Re-regulating the urban water regime in neoliberal Toronto

Douglas Young and Roger Keil

Introduction

The first observation that underlies this chapter is that all growth experienced in the Greater Toronto Area today has a significant impact on and articulation with water in its myriad forms. We will pursue this train of thought further below. Second, in as much as growth in the Toronto region is fueled by the city's role in the global economy, we can identify processes of globalization as responsible for the expansion of the urban economy and urban form. Third, as it occurs in a general climate of neoliberal public policies, growth takes place in a less regulated and market driven fashion (Keil 2002). As part of this double pressure of globalization and neoliberalization, urban water undergoes a re-regulation, whereby more marketized forms of regulation of water and land uses related to water are the norm. In Karen Bakker's terms, a shift from traditional "state hydraulic" to a "market conservation" mode of water regulation is under way in many jurisdictions (Bakker 2003b).

In Toronto, water issues have come to occupy center stage of the regional development discourse: from the waterfront to the exurban ravines of the Oak Ridges Moraine (Desfor *et al.* 2006).[1] Both development and the regulation of the region's water are now strongly influenced by the neoliberalization of Toronto's political economy. While this process has been examined in the context of urban politics (Brenner and Theodore 2002; Keil 2002; Kipfer and Keil 2002), little work has been done that links urban/regional development, water/nature and neoliberalization in Toronto. Our chapter begins to do just that and looks at the ways in which the regulation of water intersects with the growth of the city and the urban region's increased neoliberalization. How do neoliberalized regimes of urban regions intersect with water regimes?

Neoliberalization and nature in Ontario

In the Province of Ontario in the 1990s, neoliberalization of nature came first as a moderate program of ecological modernization under the social democratic NDP government (Stewart 1999); in the latter half of the decade, an

exploitative, resourcist regime took hold under the leadership of an ideologically committed group of neoconservative "common sense revolutionaries" (Ali
2004; see also McKenzie 2002; Keil 2002; Prudham 2004; Winfield and Jenish
1998).[2] Since the demise of the provincial Tory government and its replacement
by a Liberal government in 2003, a return to the more moderate policies of
ecological modernization favored by the previous NDP government has begun.
The Liberals moved quickly on a number of fronts – the preparation of a
growth management plan for a large territory referred to as the Greater
Golden Horseshoe, the creation of a large greenbelt around Greater Toronto,
and the establishment of an infrastructure funding program intended to
address what the government considers to be the province's $100 billion
dollar infrastructure deficit especially in the areas of transportation and water.
The growth management plan, *Places to Grow* (Government of Ontario 2004),
and the *Greater Golden Horseshoe Greenbelt* argue a virtuous circle of economic and population growth, preservation of open space including prime
agricultural land, and quality of life. It is too early to determine how inter-
ventionist this provincial government will be in the area of regional growth
management, however, we are inclined to believe it will not be as radical in
this regard as perhaps some environmentalist groups would like. For example,
while *Places to Grow* encourages intensification within existing urban areas, it
also accepts as given all areas currently designated in area municipalities' Official Plans for development, and goes on to suggest areas where growth beyond
those limits could take place. It is also reluctant to impose a binding plan
on area municipalities and considers legislation a last resort implementation
tool. The Greater Golden Horseshoe Greenbelt appears promising in that it
includes agricultural lands and not just "prestige" natural areas as lands worth
declaring off-limits to property developers, however, creating a greenbelt
that is physically extensive and supported by tough legislation has proven
extremely difficult given the pressure exerted by property developers geared
towards churning out tens of thousands of detached houses every year, the
apparent strong cultural preference for that form of housing, and the desire
among farmers to cash in on the development boom.

(Sub-)urbanization and water

The current discourse around land and water in Toronto is strongly influenced by real and imagined concerns about scarcity and vulnerability. While
a rabid development industry pushes on into the agricultural and forest
lands above Toronto, these violent upheavals of ways of life and ecological
systems are presented in a language of scarcity of land and resources and vulnerability of our living communities. The somewhat constructed discourse of
scarcity is precisely the foundation on which land and water markets thrive.
Clearly, the evidence that Canada in general, and Toronto in particular, are
severely compromising their land and water is there and while speaking of
"absolute" limits to growth may be overly dramatic, there is reason to assume

that the regional metabolism of the Toronto area is put to the test through continued development and increased water use. Yet, as a rhetoric of ecological modernization saturates political debates and planning discourse, the privatization drive, which has sought out water regimes as a prime target, has effectively "de-ecologized" the debate on conservation and preservation. It is essentially market efficiency and service delivery that dominate the discussion here over concerns of ecological sustainability, democracy or social justice.

The water-development nexus has had a long history in Toronto, or for that matter, Canada. When Canadian cities entered their first waves of large-scale suburbanization in the postwar decades, providing water infrastructure was seen as one of the pillars of modernization and urbanization. In fact, the explosive expansion of cities like Toronto was based on the extension of supply systems for water and sewage. We have come a long way since this first wave of – largely automobilized – suburbanization. Although, presently, development continues in leaps and bounds across the exurban landscapes, and there is talk of adding two million people to the Greater Toronto Area over the next generation, the discourse accompanying such growth has changed significantly (Keil and Graham 1998). Not only has rapid development bred its polar opposite, a strong anti-growth movement in the rural periphery, it has also changed its discursive dimension considerably: growth now seems to occur in direct and positive reference to nature and rural landscapes. In many instances, growth discourse has co-opted its critics and has taken on a "green" face: developers talk eloquently of watersheds, wetlands, ravines, ground water, and wood-lots when making their sales pitch to the masses of suburban home buyers that flood the suburban market place. And on the city's Southern development frontier, its waterfront, where 800 hectares of land have been identified for redevelopment, Lake Ontario, the potentially marshy mouths of the Don, Humber and Rouge rivers, and the body of water called "Toronto Bay" between the downtown and the city's islands, have been rediscovered as a values-producing "nature." The environment seems safely inscribed into a process of ecological modernization (Desfor and Keil 2004). Simultaneously though, technical, planning, and political modes of regulating the suburban hydrosocial cycle have been cast into crisis as subdivisions continue to eat into rural or wild lands north of the city. As exemplified by the recent struggles over development on the Oak Ridges Moraine, citizens are beginning to question the urbanization–modernization nexus which under-girded suburban residential and business development in earlier decades. The "big (sewer) pipe," once considered a sign of progress, has now become to many a symbol of the evil of sprawl (Desfor *et al.* 2006).

The urban water regime

We think all these different aspects of water have interconnections among them. They are politically, socially, economically, culturally, technologically and ecologically regulated through what can be summarized as the "Toronto water regime." This water regime is part and parcel of and sometimes – as in the present time – a core piece of the urban regime in its entirety. Inspired by urban regime analysis (Jonas and Wilson 1999; Lauria 1997), we focus on the interrelationships of the urban ecological regime, specifically the urban water regime, to the overall governing regime of the Toronto urban region in an era of neoliberalization. An urban water regime, then, entails the informal – yet also formal – institutional arrangements of private and public actors that deal with the regulation of water in an urban region. Like urban regimes, these water regimes are not islands but are intricately connected with economic, social, cultural, political, and ecological institutions, dynamics and activities on other scales than merely the local. Still, they allow us to observe concrete machinations of power, urbanization and nature in concrete terms at the urban level – they provide a distinct cut at theorizing and researching urban water in its myriad forms.

Regulation is the matter of regimes. Whether one looks at regulation as a web of social activities and economic conventions (*régulation*) or as an activity dominated by the administrative state (*règlementation*), it is the substance of the work that urban regimes do. We want to leave aside, at this point, lengthy discussions on the value of the regulation approach in its now many incarnations and currents. We value the recent discussions that have tried to re-activate regulation particularly in the context of urban ecologies, and specifically water. Bakker has discussed explicitly many of the strands of regulation theory with regards to the regulation of water and we mostly concur with her (Bakker 2000). We also value the work on "real" regulation, which it seems is attempting to "conceptualise regulation as a social process involving the state, operating at several geographical scales (national, regional, local), and involving various other public and private agents." Real regulation has been described as "a circuit of formulation, enactment, and interpretation" (Cocklin and Blunden 1998). For reasons we have no space here to explain in detail, we prefer regulation theory over "real" regulation, as it allows us to work more dynamically with other explanatory factors than the ones that are used in "real" regulation analysis. We do acknowledge, however, that "real" regulation theory has a potentially strong handle on processes of socio-natural regulation. We particularly share Cocklin and Blunden's (1998) view that "[T]he sustainability discourse is part of the continual re-regulation of society, economy and environment, and consequently the (re)production of space." On the other hand, we agree with the critique Bakker has levied against "real" regulation, which points to the limited scope of that theory as focused on the administrative state and as being "evasive" of the questions posed by the economic imperatives implied in capitalist re-production of space and nature (Bakker 2000). We would add to this critique the apparent imperviousness of the concept of "real" regulation to cultural, symbolic or discursive aspects of regulation.

Hence, we also agree with Bakker's insistence on the importance of the discursive side of the construction of the mode of regulation. She argues that "regulation is inherently (but by no means solely) a discursive practice, as well as an institutional framework embodying rules that define knowledge and legitimise authority" (Bakker 2000: 4). Discursive constructions of modes of regulation are never divorced from the very core of the material processes of economic activity that constitute them. We would argue, in addition, that in the area of ecological relations we are looking at here, such material processes are also of matter in the very concrete sense of the word, as they deal with organic and non-organic natural processes that are being constantly re-regulated with discursive interventions on "nature" while these material streams and processes, alive or not, never entirely cede their natural properties as they are socialized into modes of regulation. It is in this latter sense of overlapping realities of what can and what cannot be regulated successfully in the nature–society relationships that very little has been done to date. Most of the extant literature (Bakker and Cocklin-Blunden included) has only begun to look at the society–nature relationships as a problematic interface of the entire regulatory process.

At the core of any theory of regulation as it exists in the literature today remains the failure to properly include "nature." The regulation literature builds on the implicit understanding of the societal relationships with nature as part of the regulation of *societal*, and not of *natural* activities. We take issue with the regulation school's pervasive attempt to bring all aspects of human and non-human life under the conceptual regime of regulation, believing that it suffers from a specific short-sightedness with regard to the propensities of natural relations. These are being treated as mere extensions of other relations that the regulation school has traditionally occupied itself with, such as the wage relation or the relations of various economic actors in capitalist society. All of these relations can be – and have been – rationalized, for example, either in the rational choice tradition of "regime" theory, or in complex theories of systemic actor networks that are constituted through strategic selectivity of social actors that act through and against the state on various scales.

Water and privatization in Canadian cities

As Karen Bakker recalls, the first water networks in industrialized cities were often built by private companies:

> The poor had to rely on public taps, wells, rivers or, in the most desperate cases, stolen water. The terrible cholera and typhoid epidemics of the 19th century, combined with an apparent inability or lack of interest on the part of the private sector to finance universal provision, led the state to take over the business of water supply infrastructure.
>
> (Bakker 2003a: 17)

It has been feared that this era came to an end in the wake of the widespread privatization of services and the growth of global water companies since the 1980s, which have aggressively pursued a policy of marketization and privatization of water worldwide (Bakker 2003b; Bakker and Hemson 2000; Barlow and Clarke 2002). After Hamilton became the first Canadian city to privatize its water system in 1994 and others (Goderich, Halifax, Moncton) have experimented with various privatization and marketization schemes (Bakker 2003a), Toronto was considered a lucrative market and obvious target for future privatization. While many municipal privatization schemes proved, in fact, less than successful for water companies, they remain an option in a constantly expanding global water market. In addition, the terms privatization ("the shift in control from the public to the private sector"), marketization (the full regulation of water by market mechanisms), and commodification (often the restructuring of water management institutions and decision-making processes) have to be seen as referring to an entangled process of water provision, in which public and private actors, multinational corporations and individual citizens, municipal governments and trade unions may play various parts. In fact, in most cases, some mix of water governance through public utility, the private sector, and community or cooperative institutions seems not unusual. In addition, we know that the peculiarities of H_2O (in the hydrological cycle) and the particularities of "water" (in the hydrosocial cycle) make it difficult at all times to make water privatization a profitable enterprise (Bakker 2003a, 2003b). The issue that brought water to the fore of the public debate in Toronto was one, as we will argue here, in which overall concerns of justice guided a specific struggle around service provision.

Toronto

The Canadian census of 2001 confirmed what had been experientially evident to the residents of the Toronto region for many years: the outer ring of exurban municipalities in the four regional municipalities around the city have had population growth at a rate four times that of the city. Assuming that much of this growth in human numbers translates into low density, sprawling subdivisions, the tangible effects of this demographic explosion are even more pronounced than the statistics would express. At the same time, Toronto's inner city and some of its old suburban districts have also undergone significant growth in condominium and townhouse development, much of it at or near the city's 40 km waterfront.

This urban and suburban growth has put tremendous strain on the service networks of the Toronto region. There are huge shortfalls in affordable housing provision, transit and transportation infrastructure, educational infrastructure, landfill sites, etc. Although the city obviously lies on the shore of one of the largest bodies of fresh surface waters in the world, the growth it experiences has also begun to put considerable pressure on Toronto's water resources. At the same time, severe fiscal pressures have led the city to consider raiding a fund established for repair and replacement of water infrastructure for other municipal purposes (Moloney 2003).

Water, development and growth in the Greater Toronto Area

Overall, a new water regime is in the making in Toronto, which does – in more or less coherent fashion – create a re-structured relationship of the city with nature in this urban region. In just over 200 years Toronto has grown from a colonial outpost planted by the British in the "wilderness" to an urban region of more than five million people. Its water regime has likewise grown and adapted to changing circumstances. Progressive Era public health reforms of the early twentieth century established the universal provision of municipal drinking water and sewage treatment as a prerequisite of modernization. Throughout the nineteenth and twentieth centuries the city expanded inland from the lake in concentric circles and along water and sewage lines. Where the early twentieth century can be seen as a triumph of the discourse of *public health*, the mid and late twentieth century can be seen as a triumph of the discourse of *technology*. With the problem of waterborne disease effectively under control, the focus shifted to the technologically based steady increase in the supply of water and volume of sewage treated. The scale of the human/nature articulation related to water increased in quantum leaps from the first water treatment plant built in the nineteenth century on Toronto Island to the R.C. Harris filtration plant of the mid-1900s to the 1970s York–Durham sewer that opened a vast terrain to the north and east of the city to development. The urban explosion of the 1990s, fueled by economic growth and relaxed political and planning controls, and the current anticipation of continued regional population growth of 100,000 per year, have led to a renewed discussion of expanding supply.

At the same time though there is at present increased attention to ecological matters and the debate on growth has moved to center stage. In suburban areas the focus has been on the Oak Ridges Moraine, a band of hills that stretch 160 km from west to east across the Greater Toronto Area. Dozens of rivers and creeks have their headwaters in the Moraine, which has become mythologized as the source of water in its purest form. Citizen groups from municipalities throughout the region mounted a several years long campaign to protect the Moraine from house builders. In November 2001, the province of Ontario announced that it would implement a land use plan that would declare much of the Moraine off-limits to developers. On the one hand this can be seen as a victory for "the environment" and for water in particular, and also a victory for movement politics that ignore municipal boundaries and unite urban, suburban and rural dwellers. On the other hand perhaps it perpetuates business as usual everywhere off the Moraine. The "pure" waters of the Moraine are saved from house builders while the city of degraded nature and its filthy lake to the south are written off as unsalvageable. In between the pure and the defiled (the Moraine and the city/lake) agricultural land continues to be consumed by sprawl. While the provincial government has created a large greenbelt in the Greater Toronto Area, its efficacy in encouraging more compact development remains to be seen.

The view from the center: Toronto

In 2002, administrators and politicians in the City of Toronto, which supplies water to its own city citizens and to surrounding suburban municipalities, were accused by critics of intending to create the conditions for the privatization of water services by changing the water governance regime. At that time, a municipal proposal was on the table to establish a Toronto Water Board, fashioned along the lines of a so-called Municipal Service Board model. A previous in-house study that had floated various options of how to proceed had come under fire from the community, and amendments and changes were made as a consequence of public hearings to arrive at just this one model of the future regulation of water in Toronto.[3] Critics of the proposed Water Board maintained that it would have constituted the first step towards farming out water services and delivery to private firms. The establishment of a "commission" or "corporation" headed by a "board" of appointed rather than elected representatives is considered a major inroad to privatization. Critics further cite lack of accountability, cronyism, and the right of the board to issue 20-year contracts (possibly to private companies) as major problems of the plan. They also raised the specters of tax and water rate increases and less commitment to infrastructure investment as possible outcomes of the proposed re-regulation of water in Toronto.[4] The main thrust of the anti-privatization campaign had been brought forward by an organization called Water Watch, a loose coalition of the Toronto Civic Employees Union – Local 416, Canadian Union of Public Employees – Local 79, Canadian Environmental Law Association (CELA), Toronto-Central Ontario Construction Trades Council, the Council of Canadians, the Toronto and York Region Labour Council, the Metro Network for Social Justice, and the Toronto Environmental Alliance (TEA) (www.riversides.org/newwin/WaterWatch/aboutus.html). Water Watch turned out to be a remarkable scale-bending coalition of social justice and environmental groups. It included local labor organizations, the city's most important progressive environmental group (TEA), regional labor councils and social justice groups as well as nationally active groups such as CELA and internationally active organizations such as the Council of Canadians. In as far as this group represented social justice and environmental activists as well as labor in the form of the public service unions that represented water workers, the main bureaucratic operatives who had much of the collective expert knowledge on water in the city were on side with the groups that organized against change. Elsewhere, Debanné and Keil have argued that this made this struggle an environmental justice fight, as the main tenets of welfare statism and social solidarity were being defended and their demise would have meant widespread social and environmental injustices (Debbané and Keil 2004).

The opposition to the threatened changes to Toronto's water regime did not just consist of organized groups of progressive activists and professionals and their allies. There was widespread public support for the positions taken by Water Watch. A public opinion survey conducted for Water Watch found that "[r]esidents of Toronto are unequivocal when it comes to the management of the Toronto water system. In overwhelming proportions, and across all regions, they endorse public control over any forms of private management or contracting out" (Strategic Communications Inc. 2002). Clear endorsement of the status quo was expressed by respondents who also were found to be "overwhelmingly opposed to both the substance and the process of City Council's current deliberations over potential changes to the management of Toronto's water system" (ibid.).

In addition to the attempted restructuring of water regulation in the City, huge lands adjacent to waterfronts and rivers are being brought under new forms of private control and more flexible and unaccountable forms of management including simplified approval processes for development (Bunce and Young 2004). The ensuing (sub)urbanization-water complex is the subject of large-scale efforts to redefine the Waterfront. A City document called *Making Waves: Principles for Building Toronto's Waterfront* (City of Toronto 2001) describes a new aquatic future "in purple prose thick with water metaphors" (Bunce and Young, op. cit.): "[t]he benefits that will ripple out from a revitalized Central Waterfront will extend beyond its boundaries and will wash across the whole of the city" (City of Toronto op. cit.: 13; Bunce and Young op. cit.: 217).

Conclusion

The regulatory regime of urban water in Ontario has suffered a few major blows in recent years. Particularly the water scandal of Walkerton, a rural community northwest of Toronto, where 7 were killed and thousands fell ill from E. coli in May of 2000, threw provincial water policy into deepest crisis. An independent fact-finding task force reported in the spring of 2002 that the regulatory regime undermined by the Provincial government had failed not just the people of Walkerton but had cast a shadow over the entire system of water supply in the province. A Safe Drinking Water Act, brought in late October 2002 by the Tory government, took up some of the recommendations of the Provincial O'Connor Report, particularly those that referred to the "pumps and pipes" part of the system, but left unaddressed the larger ecological and social question of source protection. The regulation of the Toronto water regime has recently shown some pressure points both at the center and at the edges. The previous Mayor and more market-oriented councilors as well as some city staff championed a review of the city's water governance system. The outcome was a recommendation to create a Toronto Water Board, a governing body, which would consist of mostly appointed members. The battle lines in this struggle are clearly drawn. Residents of the City of Toronto opposed the potential privatization of their water supply and delivery system and successfully fended off what were considered preliminary steps in that direction. It is understood that any change to the public status of water supply, demand management and delivery would seriously endanger the safety and affordable availability of drinking water in Toronto. While the struggle over possible privatization was underway, the city's planners once again stepped up their efforts to build their city on the merit of the symbolic and aesthetic qualities of water, particularly in the form of its vacant land alongside the shore of Lake Ontario. In the meantime, exurbanization rages on with a development industry largely unchecked by government regulation and an active political endorsement of the continuation of large-scale, technologically conventional systems of water supply and reticulation as well as sewage systems that allow the continued overflow of the suburban ring into ever further copies of itself. At the same time, though, citizens in the outer suburbs of Toronto have begun to fight the degradation of their – exclusive – living arrangements at the city's edge.

Neoliberalizing nature encounters all manner of contradictions: the general terrain is neoliberal but the outcomes of social struggles are contingent. In these struggles over water and land, important new connections of city and nature are being forged. Democratic discourse and sustainable development are open-ended processes with much continued conflict and contradiction even once more democratic and more ecological developments have been put in motion.

Acknowledgments

A longer version of this chapter was Young, D. and Keil, R. (2005) "Toronto Water: A Political Ecology," *Capitalism, Nature, Socialism* 16(2): 61–83. Research for this chapter was supported by a small grant from the Faculty of Environmental Studies, York University, Toronto.

Notes

1 The Oak Ridges Moraine is a headwaters region to the north of Toronto that is under intense development pressure.
2 In fairness, it needs to be pointed out that under pressure of their exurban constituencies, the Ontario Tories made some strides towards preservations of the Oak Ridges Moraine.
3 Personal communication, Eduardo Sousa.
4 Information was taken from various smaller publications of the organization Waterwatch. Please see www.torwaterwatch.org for sources.

References

Ali, S. A. (2004) "A Socio-Ecological Autopsy of the *E. coli* O157:H7 Outbreak in Walkerton, Ontario, Canada," *Social Science and Medicine* 58: 2601–12.

Bakker, K. (2000) "Privatising Water, Producing Scarcity: The Yorkshire Drought of 1995," *Economic Geography* 76: 1.

—— (2003a) "Liquid Assets," *Alternatives* 29(2): 17–21.

—— (2003b) "A Political Ecology of Water Privatization," *Studies in Political Economy* 70: 35–58.

Bakker, K. and Hemson, D. (2000) "Privatizing Water: Hydropolitics in the New South Africa," *South African Journal of Geography* January.

Barlow, M. and Clarke, T. (2002) *Blue Gold: The Battle Against Corporate Theft of the World's Water*, Toronto: Stoddart.

Brenner, N. and Theodore, N. (eds.) (2002) *Spaces of Neoliberalism*, Oxford: Blackwell.

Bunce, S. and Young, D. (2004) "Image-Making by the Water: Global City Dreams and the Ecology of Exclusion," in International Network of Urban Research and Action, Raffaele Paloscia (ed.) *The Contested Metropolis: Six Cities at the Beginning of the 21st Century*, Basel: Birkhäuser.

City of Toronto (2001) *Making Waves: Principles for Building Toronto's Waterfront*, Toronto: Toronto City Council.

Cocklin, C. and Blunden, G. (1998) "Sustainability, Water Resources and Regulation," *Geoforum* 29(1): 51–68.

Debanné, A. and Keil. R. (2004) "Multiple Disconnections: Environmental Justice and Urban Water in Canada and South Africa," *Space and Polity* 8: 2:

Desfor, G., Keil, R., Kipfer, S. and Wekerle, G. (2006) "From Surf to Turf: No Limits to Growth in Toronto," *Studies in Political Economy*.

Government of Ontario (2004) *Places to Grow: Better Choices. Brighter Future: A Growth Plan for the Greater Golden Horseshoe*. Discussion paper, Summer, Ontario.

Jonas, A. and Wilson, D. (eds.) (1999) *The Urban Growth Machine: Two Decades Later*, Albany, NY: The State University of New York Press.

Keil, R. (2002) "'Common Sense Neoliberalism': Progressive Conservative Urbanism in Toronto, Canada," *Antipode* 34(3): 578–601.

Keil, R. and Graham, J. (1998) "Reasserting Nature: Constructing Urban Environments after Fordism," in B. Braun and N. Castree (eds.), *Remaking Reality: Nature at the Millennium*, London and New York: Routledge.

Kipfer, S. and Keil, R. (2002) "Toronto Inc? Planning the Competitive City in the New Toronto," *Antipode* 34(2): 227–64.

Lauria, M. (1997) *Reconstructing Regime Theory: Regulating Urban Politics in a Global Economy*, London: Sage Publications.

McKenzie, J. I. (2002) *Environmental Politics in Canada*, Toronto: Oxford University Press.

Moloney, P. (2003) "Council Seeks to Tap Water Fund," *Toronto Star*, February 26, p. B3.

Prudham, S. (2004) "Poisoning the Well: Neoliberalism and the Contamination of Municipal Water in Walkerton, Ontario," *Geoforum* 35: 343–59.

Stewart, K. (1999) "Greening Social Democracy? Ecological Modernization and the Ontario NDP," PhD dissertation, Department of Political Science, York University, Toronto.

Strategic Communications Inc. (2002) *Managing Toronto's Water: A Public Opinion Poll Commissioned by Toronto Water Watch* (Toronto: Water Watch), available at: www.riversides.org/newwin/WaterWatch/aboutus.html

Winfield, M. S. and Jenish, G. (1998) "Ontario's Environment and the 'Common Sense Revolution,'" *Studies in Political Economy* 57(Autumn): 129–47.

Part II

Commentary

12 Neoliberalism and the regulation of "environment"

Neil Brenner and Nik Theodore

The concept of neoliberalism has been widely used to characterize the resurgence of market-oriented institutional shifts and policy realignments across the world economy during the post-1980s period (see, for example, Bourdieu 1998; Gill 1998). Technically, neoliberalism refers to a set of doctrines regarding the appropriate framework for economic regulation. More recently, however, the term has been appropriated by scholars and activists to describe the institutional, political and ideological reorganization of capitalism that has been imposed through the attempted institutionalization of "free market" doctrines since the global economic crises of the mid-1970s.

For the most part, the "great transformations" (Polanyi 1957) associated with neoliberalism have been investigated with reference to national regulatory trends (for instance, the rise of Reaganism in the USA and Thatcherism in the UK) and supranational institutional realignments (for instance, the role of the World Bank and the IMF in imposing structural adjustment programs upon developing countries). Since the late 1990s, however, an impressive body of work has been generated by scholars who have reflected in some detail upon the variegated *geographies* of neoliberal restructuring. These encompass diverse, large-scale sociospatial transformations – of political-economic structures, of urban-regional development patterns, of regulatory arrangements and of interscalar interactions (for an overview, see Peck 2003, 2001). As the contributions to this book illustrate, this line of research has also begun to explore the question of how neoliberalization processes have transformed society/nature interactions at various scales and in diverse geographical settings.

The new scholarship on the geographies of neoliberalism has generated a number of fruitful insights that have significant implications for empirical research (Brenner and Theodore 2002a, 2002b; Peck and Tickell 2002; Tickell and Peck 2002). For present purposes, we offer a series of brief propositions that is intended to capture some of the key ideas developed in this emergent literature:

1 *Neoliberalism is a process.* Neoliberalism is not a fixed end-state or condition. Rather, it represents a *process* of market-driven social and spatial transformation ("neoliberalization").

2 *Neoliberalism is articulated through contextually specific strategies.* Neoliberalism does not exist in a single, "pure" form. Rather, it is always articulated through historically and geographically specific strategies of institutional transformation and ideological rearticulation.

3 *Neoliberalism hinges upon the active mobilization of state power.* Neoliberalism does not entail simply the "rolling back" of state regulation and the "rolling forward" of the market. Instead, it generates a complex reconstitution of state/economy relations in which state institutions are actively mobilized to promote market-based regulatory arrangements and to extend the process of commodification.

4 *Neoliberalization generates path-dependent outcomes.* Neoliberalism does not engender identical (economic, political or spatial) outcomes in each context in which it is imposed. Rather, as place-, territory- and scale-specific neoliberal projects collide with inherited regulatory landscapes, contextually specific pathways of institutional reorganization crystallize that reflect the legacies of earlier modes of regulation and forms of contestation.

5 *Neoliberalization is intensely contested.* Neoliberalization, understood as the attempt to extend the process of commodification through the imposition of market-based regulatory arrangements and sociocultural norms, is aggressively contested. It is opposed by diverse social forces concerned to preserve non-market or "socialized" forms of coordination that constrain unfettered capital accumulation and impose limits upon the process of commodification.

6 *Neoliberalization exacerbates regulatory failure.* The imposition of neoliberalism has not established a framework for sustainable development, stable political regulation or social cohesion. Rather, neoliberalization projects are deeply contradictory insofar as they tend to undermine many of the economic, institutional and geographical preconditions for socioeconomic revitalization. Thus, instead of resolving the political-economic crisis tendencies of contemporary capitalism, neoliberalism exacerbates them by engendering various forms of market failure, state failure and governance failure (Jessop 1998).

7 *The project of neoliberalism continues to evolve.* The failures of neoliberalism have not triggered its abandonment or dissolution as a project of radical institutional transformation. To the contrary, this project has continued to reinvent itself – politically, organizationally, spatially – in close conjunction with its pervasively dysfunctional social consequences.

The operationalization of these propositions in the context of concrete, empirical research on any aspect of contemporary capitalism presents significant methodological challenges. In our own work, we have coined the term "actually existing neoliberalism" in order to underscore the profound disjuncture between orthodox neoliberal ideology and the complex, contested and uneven geographies of regulatory change that have emerged in and through projects of neoliberalization (Brenner and Theodore 2002a). In addition, the concept of actually existing neoliberalism is intended to demarcate a terrain for critical inquiry into the contextually specific pathways of neoliberalization that are crystallizing in cities, regions and states throughout the world economy.

The chapters included in this part of *Neoliberal Environments* do not explicitly grapple with the conceptualization of neoliberalization processes, but they do fruitfully extend our understanding of such processes by exploring their ramifications in the field of environmental governance. Specifically, the chapters by Bakker, Robertson, Hollander and Young and Keil investigate the ways in which neoliberal projects of marketization and commodification have been imposed upon particular aspects of society/ nature relations in diverse geographical settings – including the regulation of water production and consumption in England, Wales and Toronto; the regulation of wetland land-use patterns in exurban Chicago; and the regulation of agricultural production in rural Florida. Since Polanyi's (1957) classic analysis of the "great transformation" associated with the attempt to create "self-regulating markets" in the late nineteenth and early twentieth centuries, critical scholars have recognized that (a) the operation of markets is not self-sustaining, but hinges upon the construction and maintenance of regulatory arrangements; and (b) the commodification of social life is not the natural "order of things," but can be accomplished only provisionally, through the disciplining impacts of market-oriented institutional structures and rule-regimes. Each of the chapters included in this part reinforces and illustrates these contentions. Nature – whether manifested in the form of water, land or agricultural produce – is not, in itself, a commodity; yet, it may be subjected to a *logic* of commodification insofar as it is appropriated according to institutionalized principles of exchange, private ownership and profitability. Concomitantly, the contributions in this part reveal the ways in which strategies to subject nature to the logic of the commodity may generate unintended, and deeply dysfunctional, outcomes. For, under neoliberal rule-regimes, water and food may not be distributed equitably even when they are abundant; natural landscapes may be degraded through overuse or inadequate protection; and social needs may be neglected due to the private appropriation of collective natural resources. Contrary to the claims of neoliberal pundits, such "externalities" are not the result of inadequate or insufficient marketization, but are intrinsic to the very workings of capitalist market economies (Polanyi 1957; Gill 1998). As the chapters under discussion here show, the disruptive consequences of such "illogics" are severely exacerbated under neoliberal regulatory arrangements.

Above and beyond these general insights into the logics and illogics of commodification, the case studies presented in Part III of the book also provide useful insights into what we might term the "spatial selectivity" (see Jones 1997; Brenner 2004) of neoliberalism as a political strategy. For, as all the chapters illustrate, the impacts of neoliberal approaches to environmental regulation are not distributed uniformly across the landscape and do not engender a smooth locational surface on which markets can self-regulate. Rather, these policies have differentially impacted locations, places, territories and scales across the globe; their geographical ramifications are thus deeply variegated.

- Bakker's chapter, for example, explores the geographies of "market environmentalism" associated with neoliberal approaches to resource management – specifically, water supply – in England and Wales. In investigating the privatization of the water industry in these regions, Bakker emphasizes the inherent difficulties in commodifying water, whose "continually circulating, scale-linking qualities and biophysical, spatial, and socio-cultural characteristics render it particularly resistant to commodification" (p. 111). For this reason, neoliberal projects of deregulation were swiftly met with diverse, reregulatory strategies intended to manage some of the dislocations and failures associated with privatization strategies. Crucially, however, Bakker underscores that the shift from a state-hydraulic model of water management to a neoliberalized model also entailed a significant redistribution of socio-environmental costs and burdens. Whereas the state-hydraulic model promoted redistribution but engendered extensive environmental destruction, the neoliberal model actually improved water quality but created differential levels of service provision and access among consumers. Neoliberalization, in this sense, established not only a new framework for environmental governance, but also new "power geometries" in and through which access to this basic resource has been controlled – and contested.
- Robertson's chapter, which explores the problem of habitat protection in the wetlands of exurban Chicago, illuminates a different aspect of the geographies of environmental regulation. Here, in an environment under intense pressure due to accelerating urban sprawl, a market-based model for wetlands preservation was established. This permitted large-scale developers to engage in selected forms of construction within environmentally protected areas. Through a specific form of wetlands "banking," neoliberal technologies of measurement and benchmarking were harnessed in an effort to create well-functioning markets. However, Robertson's chapter underscores the deeply conflict-ridden character of these efforts – the different constituencies involved in the transactions deploy only partially commensurable systems of measurement, and this leads to considerable insecurity regarding the trajectory of the accumulation process in these areas. For Robertson, therefore, the attempt to neoliberalize environmental governance poses deep epistemological challenges at once for land developers and for all other social forces concerned to influence the use and appropriation of the wetlands.
- In her chapter, Hollander shifts the focus to international trade relations and their impact upon the governance of agricultural production in a

particular region. Her account of the rather elusive notion of "multi-functionality" traces the collision between a neoliberalized international rule-regime (embodied in the agricultural policies of the European Union and the trade regulations of the GATT/WTO regime) and the local spaces of agricultural production in the Everglades Agricultural area of southern Florida. As Hollander's account shows, this collision has generated a host of new agro-environmental problems that affect natural habitats, economic development and everyday livelihoods throughout the "Sugar Bowl" region. The project of multifunctionality does not resolve these regulatory problems, but repositions them in a new geo-regulatory, politico-cultural context. Faced with emerging environmental mandates aimed at maintaining long-run environmental sustainability and biodiversity in the region, new lines of conflict emerge among agribusiness corporations, family farmers, environmental activist organizations, urban residents and other local constituencies. Thus, even when combined with ostensibly "progressive" socio-environmental agendas, the process of neoliberalization appears to intensify, rather than alleviate, the constitutive unevenness of capital circulation, and associated attempts to regulate its endemic contradictions. Neoliberalization, in short, enhances geographical differentiation at all spatial scales and in all settings in which it is mobilized.

- Finally, the chapter by Young and Keil returns to the theme of water management by investigating the politics and discourses of water privatization in the Toronto metropolitan area. Their analysis embeds the real and imagined geographies of water regulation within the broader process of urban and regional development and political-institutional restructuring in Ontario. A new Toronto water management regime emerged, they argue, at the contested interface between Liberal proposals to promote "ecological modernization" and new discourses regarding water scarcity and the "limits to growth" in the face of accelerating urban sprawl. In this manner, Young and Keil show how the appropriation and distribution of water are intertwined not only with evolving (state) institutional structures but also with the broader processes of "creative destruction" associated with capitalist urbanization in a globalizing city. Their account surveys various social, political and geographical perspectives on the politics of water regulation in Toronto, underscoring the deeply contested character of attempts to privatize access to this essential resource. While Bakker's account of water privatization emphasized the role of national and regional political institutions in England and Wales, Young and Keil focus most directly on the urban scale, where the politics of water is closely intertwined with broader struggles regarding planning and diverse citizens' initiatives. For Young and Keil, then, the unevenness of neoliberalized forms of environmental governance stems in significant measure from the conflicts they provoke, at once in neighborhoods, cities and regions. And yet, while they insist that neoliberalization processes are inherently contradictory, they underscore the fundamentally contingent – that is, open-ended – character of the struggles that flow from these contradictions.

Taken together, the contributions included in this part reveal in powerful detail the constitutively uneven character of neoliberal strategies of environmental regulation: they are not associated with a singular project of sociospatial transformation; they do not engender homogeneous geographical outcomes; they do not create a stable, self-regulating framework of society/nature relations; and they produce all manner of unintended socioenvironmental dislocations. For, within each national, regional and local context, neoliberal approaches to environmental governance are mobilized in contextually specific forms, and the latter in turn interact with inherited society/nature configurations in highly variegated, destabilizing and often unpredictable, ways.

These considerations in turn suggest that purely theoretical work on neoliberal forms of environmental governance – and, more generally, on the geographies of neoliberalism as a whole – contains basic limitations. While the ideologies of neoliberalism can be productively deconstructed at an abstract-theoretical level, it is only through concrete complex research – guided, of course, by theory – that the "lean and mean" geographies of actually existing neoliberalism can be deciphered "on the ground." At the present time, it is possible to advance only the most general propositions, such as those proposed above, regarding the nature of these geographies. Against this background, the concrete investigations provided in this book are of particular analytical value, for they point towards a broader research agenda on the transformative *strategies,* restructuring *pathways* and sociospatial *dislocations* associated with neoliberalization processes. It is to be hoped that further research along these lines will enable scholars not only to deepen their theoretical grasp of neoliberalization processes, but also to extend and differentiate their understanding of their highly variegated origins, manifestations and consequences. For the moment, though, our understanding of how "market rule" is implicated in contemporary patterns of sociospatial and society/nature transformation remains seriously incomplete.

References

Brenner, N. (2004) *New State Spaces: Urban Governance and the Rescaling of Statehood,* New York: Oxford University Press.

Brenner, N. and Theodore, N. (2002a) "Cities and the Geographies of 'Actually Existing Neoliberalism,'" *Antipode* 33(3): 349–79.

—— (eds.) (2002b) *Spaces of Neoliberalism: Urban Restructuring in North America and Western Europe,* Oxford: Blackwell.

Bourdieu, P. (1998) *Acts of Resistance: Against the Tyranny of the Market,* New York: New Press.

Gill, S. (1998) "New Constitutionalism, Democratisation and Global Political Economy," *Pacifica Review* 10(1): 23–38.

Jessop, B. (1998) "The Rise of Governance and the Risks of Failure: The Case of Economic Development," *International Social Science Journal* 155: 29–46.

Jones, M. (1997) "Spatial Selectivity of the State? The Regulationist Enigma and Local Struggles over Economic Governance," *Environment and Planning A* 29: 831–64.

Peck, J. (2001) "Neoliberalizing States: Thin Policies/Hard Outcomes," *Progress in Human Geography* 25(3): 445–55.

—— (2003) "Geography and Public Policy: Mapping the Penal State," *Progress in Human Geography* 27(2): 222–32.

Peck, J. and Tickell, A. (2002) "Neoliberalizing Space," in N. Brenner and N. Theodore (eds.) *Spaces of Neoliberalism: Urban Restructuring in Western Europe and North America*, Oxford: Blackwell, pp. 33–57.

Polanyi, K. (1957) *The Great Transformation*, Boston, MA: Beacon Press.

Tickell, A. and Peck, J. (2002) "Making Global Rules: Globalization or Neoliberalization?" In J. Peck and H. Wai-chung Yeung (eds.) *Remaking the Global Economy: Economic-Geographical Perspectives*, London: Sage, pp. 163–81.

Part III

Devolution and neoliberal governmentalities

13 Poisoning the well

Neoliberalism and the contamination of municipal water in Walkerton, Ontario

Scott Prudham

Irresponsibility is the organizing principle of the neo-liberal vision
(Günter Grass, "The Progressive Restoration: A Franco-German Dialog,"
New Left Review 14, March/April 2002, p. 71)

Introduction: poison in the water

For residents of Walkerton, Ontario, the Victoria Day weekend of 2000 began as had many before it. Viewed as the start of summer, Victoria Day (one week prior to the American Memorial Day Holiday) offers Canadians an opportunity to break out the barbeque, open up cottages, air out tents, visit friends and family, and talk about playoff hockey. In Walkerton, a spate of thunderstorms in the week preceding the holiday did little to dampen enthusiasm for an annual rite. Indeed, as the weekend arrived, though warning signs had already appeared, there was little hint of an imminent calamity. But by Monday morning, Walkerton's first resident had died from drinking poisoned town water. The death of Lenore Al would be followed by six more. Despite a boil-water advisory issued by the region's Medical Officer of Health on Sunday May, 21st, in excess of 2,300 area residents became infected; many survivors suffered seriously, and continue to experience long-term effects both physical and psychological.[1]

The proximate cause of the infections and deaths was soon apparent: contamination of treated municipal water by *Escherichia coli* and *Campylobacter jejuni* bacteria. A particularly deadly strain of E. coli known as O157:H7 and found in the stomachs of cattle was implicated in the most severe cases, and in all of the deaths.[2] Yet, where had the bacteria come from? How did they get into the town's water? And why were they found in *treated* water, despite chlorination systems, testing procedures, and claims by Ontario government authorities that provincial regulations safeguarding Ontario drinking water were adequate? In a nation routinely ranked at or near the top of the United Nation's Human Development Index, a scandal quickly erupted over how what had come to be taken so utterly for granted – the provision of safe municipal drinking water – could fail so catastrophically.

The provincial government, seeking to distance itself from the tragedy, consistently portrayed the incident as a combination of ostensibly freak "natural" circumstances and administrative bungling by water utility managers. Yet, in this chapter, I examine the Walkerton tragedy instead as a kind of "normal accident" (Perrow 1999), a case of what Jamie Peck (2001) has termed neoliberalism's "thin policies and hard outcomes." I argue that the Walkerton incident implicates in particular neoliberal reforms visited in the wake of the Ontario elections of 1995. Under the auspices of what was billed as a "Common Sense Revolution," sweeping, unprecedented and highly ideologically charged changes were introduced in Ontario bearing many of the familiar hallmarks of so-called "rollback neoliberalism" (Peck 2001; Peck and Tickell 2002). Critically, neoliberalization in Ontario was predicated in significant measure on the re-configuration of provincial environmental governance and in ways that contributed to the Walkerton incident. Sweeping rollbacks cut a broad swath through Ontario's environmental regulatory apparatus, undermining the capacity of regulatory agencies, placing a marked "chill" on setting and enforcing standards, and creating specific regulatory gaps. Shifts in agricultural and water quality regulation in particular induced failures of oversight and accountability that helped produce the conditions for regulatory failure, helping to "produce" the Walkerton tragedy in identifiable ways.

The chapter is organized in the following manner. The first section provides some brief elaborations on the notion of a normal accident produced by neoliberal governance reforms. The subsequent section addresses locally specific factors that contributed to the poisoning of Walkerton's water, including the conjunction of a distinct hydrological regime, local and regional livestock production, and the practices of municipal utility officials in (mis)managing the town's water supply. Subsequently, the chapter turns to chronicling the establishment of organized "irresponsibility" (to borrow Grass's apt notion, quoted above) in environmental governance via neoliberal reforms introduced subsequent to the 1995 Ontario elections.

Confronting the production of environmental risk under neoliberalism

As contributions to this volume attest, neoliberalism, or better, specific "neoliberalizations" represent significant sources of restructuring in socionatural relations. This chapter focuses on the role of neoliberal reforms in Ontario, Canada, in generating new environmental risks. I stress that neoliberalization in Ontario is implicated in the "production" (Smith 1984) of new environmental hazards, and that locating the Walkerton tragedy in relation to neoliberal governance reforms allows this incident to be understood as a "normal" accident of neoliberalism. The concept of a normal accident was developed by Charles Perrow (1999) as a way of describing catastrophic failures in systems whose characteristics make such events inevitable. To be fair, Perrow did not have neoliberalism in mind, and his notion is intended to be applied to emergent properties in technologically and organizationally elaborate systems (e.g. nuclear power production). I adopt and adapt the term here to denote the ways in which *organized irresponsibility* is built into regulatory systems. Drawing on Perrow, this view stresses that while the actual circumstances of "accidents" are indeed important, so too is the broader political economy of neoliberal regulatory reforms shaping both the probabilities that accidents will occur, and the likely consequences when they do.

There are parallels to be drawn with Ulrich Beck's widely influential thesis on environmental risk and a new so-called "Risk Society" (Beck 1999; Beck and Ritter 1992). Beck argues that environmental risks are becoming endemic and pandemic in late modern society, as are the politics and institutional strategies comprising sociological responses. He specifically suggests that exposure to new sources of environmental risk cuts across traditional class fractions to "produce" a new politics of risk, a perspective that at first glance would seem directly applicable to the contamination of municipal water systems (since, ostensibly, we all drink the same water; whether this is true or not is perhaps better addressed elsewhere in the volume). In this light, Walkerton might seem a particularly poignant marker of the new era of environmental risk and anxiety, an extreme example of more generalized phenomena. But I am reluctant to embrace this narrative fully, as much as Beck's attention to the sociological importance of environmental risk is necessary. This is because Beck's original thesis suffers from a diffuse account of the politics of risk production. Thus, as Ted Benton (1997) points out, while Beck's thesis highlights the significance of new environmental risks and their politics, his theory tends to downplay the specific political economies of such risks, making them seem endemic to a highly generalized late modernity. Beck's Risk Society thesis makes it difficult to locate anything as particular as regulatory restructuring and political struggles over the regulatory apparatus of capitalist states as significant sources of risk production. Yet, as Benton demonstrates by drawing on the links between deregulated animal feed production in Thatcherite Britain and the UK's mad cow disease outbreak, new environmental risks are often closely tied to struggles over state regulation of private capital and market allocation. And this is one of the central features of neoliberalism: political struggle over the role of states in mediating the power of private decision making and market allocation vis-à-vis the environment.

There is in this the echo of a prominent theme in the political ecology literature, namely that local, ecologically specific and even seemingly unique events, processes, and crises tend to take place within a broader political economic context that helps to produce them in observable ways, as well as to shape their consequences. Seeking out this broader context, as I try to do in this chapter, points toward state-centred and neoliberal-inspired shifts in the social regulation of nature in Ontario as germane to any explanation of what happened in Walkerton (however over-determined causation may otherwise be). In this, the account offered here also echoes one of Karl Polanyi's (1944) most insightful aphorisms in reference to an earlier generation of economic liberalism to create what he called market self-regulation; "*laissez-faire* is planned." And as Polanyi also stressed, the socio-ecological consequences of this planning are discernible and potentially disastrous.

Walkerton

Walkerton is a pastoral town of approximately 5,000 people located in southern Bruce County, in the heart of central-southern Ontario, approximately 150 km north-west of Toronto (see Figure 13.1). Set in the rolling countryside of the upper Saugeen River watershed, it acts as the administrative seat for Bruce County, and has predominantly served as a kind of commercial and service hub for the surrounding, predominantly rural and agricultural area.

Karst

In terms of local context, the particular karst hydro-geology of the Walkerton area is extremely significant to understanding why the tragedy occurred where it did. "Karst is terrain with distinctive hydrology and landforms arising from a combination of high rock solubility and well developed secondary porosity" (Ford and Williams 1989: 1). Karst formations, most of which exist in carbonate rock, account for about 7–12 percent of the earth's

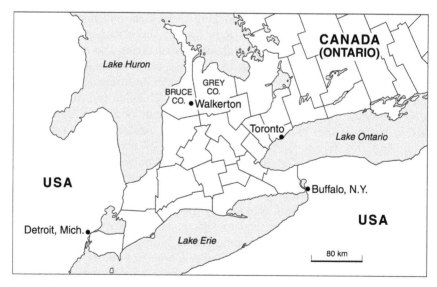

Figure 13.1 Walkerton, Ontario.

surface, yet provide water for as much as one quarter of the world's population (Drew, Hötzl *et al.* 1999). Distinct features of karst formations include networks of channels and pockets in the rock, providing avenues for underground water transport and storage.

Unlike most of southern Ontario, the carbonate bedrock in the Walkerton area is quite close to the surface. Despite relatively few visible superficial indications, limestone and dolostone formations underlie an extremely thin surface layer of gravel and soil, typically on the order of 5–15 m deep (Worthington, Smart *et al.* 2001). Because of its proximity to the surface, this layer of bedrock in the Walkerton area comprises an hydrologically active form of karst (Cowell and Ford 1980; 1983; Drew, Hötzl *et al.* 1999). Of particular concern in this context is the potential for runoff contaminated with bacteria (including E. coli and *Campylobacter jejuni*) to seep into groundwater and be propagated into wells and springs used for drinking water. This appears to be what happened in the Walkerton case.

In particular, one of the town's wells known simply as Well 5 has been implicated as the source of most if not all of the contamination during May of 2000 (O'Connor 2002a). Following significant rain events, during which runoff is elevated, tests were conducted to determine the association between peak fecal coliform contamination in the well, and peak runoff (see Worthington *et al.* 2001). The results indicated a clear association, with lag times on the order of 1–4 days. This strongly suggests that the source of contamination in Well 5 was contaminated surface water draining into the aquifer and contaminating the well. This connection is strengthened by evidence that Walkerton's water intake was contaminated on the order of 5–7 times per year preceding May of 2000, usually in the aftermath of significant rainfall.

Stan and Frank Koebel

Like 80 percent of the municipal water systems in Ontario, Walkerton's drinking water system and supply is managed by a public utility chartered and funded in part by the province (C.N. Watson and Associates 2001). Walkerton's utility is run by a combination of elected representatives on the Walkerton PUC and staff hired by the board to run day-to-day operations.

There is ample evidence to indicate that negligence and possible criminal misconduct by two brothers, Stan and Frank Koebel (respectively, general manager and foreman at the Walkerton PUC), contributed to the Walkerton incident. On Saturday, May 13th 2000, after severe rainstorms in the area, Frank Koebel was responsible for conducting routine tests on pumping rates and chlorine residuals. The chlorine residuals test indicates the extent to which chlorine remains in the water supply following treatment. High chlorine residuals indicate low levels of contamination. Low chlorine residuals suggest bacteria in the raw well water, while an absence of chlorine indicates bacteria remain in treated water. However, on that day, as had become the custom at Walkerton's PUC, Frank simply falsified log books kept by the utility rather than actually conducting the test. Subsequently, A&L Laboratories telephoned and faxed test results to Stan Koebel indicating extensive contamination of the town's water supply. Stan Koebel, despite his position as the PUC's general manager, did nothing (O'Connor 2002a).

Subsequent investigation revealed that staff bungling in the lead-up to the Walkerton tragedy was hardly exceptional. Staff under Stan Koebel's direction routinely mislabelled samples, neglected to chlorinate drinking water altogether, submitted false reports to the MOE, and apparently made a regular practice of drinking alcohol at work. In addition, although Stan Koebel had certification as a water system operator, at no time did he complete any formal training as such. Instead, Koebel had been "grand-fathered" into compliance with new regulations introduced in the 1980s, and had thereafter received pro forma renewals of his certification. Incredibly, Koebel testified to the public inquiry in the Walkerton affair that *he had never read the province's guidelines on unsafe drinking water, and did not even know what E. coli were* (O'Connor 2002a).

Neoliberalism in Ontario

Seeking to deflect attention from its own culpability, the Premier's office immediately and consistently blamed the Walkerton tragedy on local officials, eventually bringing criminal charges against the Koebel brothers. Yet, while the conduct of PUC staff unquestionably contributed, their actions had a context. Most obviously, the Koebels and other staff were enabled to function by a regulatory system in which specific reforms introduced in a climate of neoliberalization exacerbated the risks that local incompetence could have tragic consequences. Particularly relevant are sweeping changes to the architecture of provincial environmental regulation, including the creation of gaps in the oversight of agricultural waste disposal practices, groundwater management, and municipal water administration.

The adoption of neoliberal policies in Ontario pre-dates election of a formally neo-conservative provincial administration in 1995. Specifically, in the early 1990s, the social democratic government of Ontario's New Democratic Party (NDP) led by Premier Bob Rae was confronted by deteriorating economic conditions, disintegration of the progressive coalition underpinning the provincial NDP, and extreme pressure from domestic and international finance capital to adopt neoliberal fiscal reforms. In panic, Rae's NDP embraced austerity. Provincial environmental spending suffered some of the deepest cuts, dropping to $352 million in 1994–95 from over $800 million just three years prior (Krajnc 2000). Seeking specific relief from the costs of water testing under the auspices of the MOE, the NDP also restructured provincial–municipal relations governing drinking water management. In 1993, the province for the first time introduced charges to local municipalities for the costs of water tests undertaken by the provincial MOE, in an

effort at fiscal devolution and cost recovery. In addition, the Rae administration opened the door for testing by independent, private labs.

The environment of common sense

Introduction of classically neoliberal policies (largely of the rollback variety)
by a left-of-centre provincial government in the early 1990s suggests how
far-reaching the neoliberal "consensus" in governance had by this time
extended. Even so, it was not until the election of 1995, in the midst of
continued recession, that a much more far-reaching and revanchist neoliberal project was initiated in Ontario by the provincial Progressive Conservative or "Tory" Party. The Tories, traditionally a party of patrician
conservatism, were re-invented as a neoliberal juggernaut under the leadership of Mike Harris who drew on a new, ideologically charged coalition of
suburban small business support and a fanatical contingent of young conservatives in the party (Keil 2002).

Dubbed the "Common Sense Revolution," the Harris administration
offered a familiar formula reminiscent of the Thatcher and Reagan eras
(Jessop *et al.* 1990; Peck 2001; Peck and Tickell 1992; 1995; Tickell and
Peck 1995). It was in many ways textbook rollback: steep spending cuts; tax
reduction for the wealthiest; welfare and workfare reform; sweeping state
retrenchment; and liberalization of provincial labour laws and markets (Keil
2002). "Common sense" (of course) stood for markets, markets, and more
markets, expanding the scope of private decision-making and accumulation
via three familiar tropes of rollback neoliberalism: fiscal austerity; deregulation and re-regulation; and privatization.

Critically, these strategies were visited in significant measure by deep
restructuring of environmental governance in Ontario. Among the first
items on the Common Sense agenda were cuts to environmental regulatory
and resource management agencies, with the first Common Sense budget
cutting total environmental spending by one-third from levels already
reduced under Rae administration rollbacks. This time, cuts targeted not
only discretionary spending but also core elements of environmental regulatory and administrative capacity, including the elimination of one-third
of the MOE's 2,000 total staff and the termination of 2,100 MNR jobs
(Krajnc 2000). Critically, the cuts were highly ideological in character, initiated from the political centre of government with little consideration of
whether they would compromise existing and statutory commitments
(O'Connor 2002a: 34).

In some instances, fiscal cuts were accompanied by administrative and
regulatory ones, consolidating and eliminating "loose ends" created by fiscal
rollbacks. Thus, during the spring of 1996 when an Ontario fruit and vegetable inspection programme aimed at controlling pesticide residues and conducted by the Ministry of Agriculture, Food, and Rural Affairs (OMAFRA)
became impossible to execute due to funding and staff cuts, the programme

was simply terminated altogether. In addition, however, direct, targeted rollout style environmental re-regulation led to broad, often seemingly unrelated changes to a host of laws and administrative procedures. Bill 26 (known simply as the Omnibus Bill), for example, amended an amazing 44 different statutes all at the same time. The overall flavour of the Bill emphasized industry self-regulation, and the replacement of mandatory with voluntary standards and participation, including elimination of provincial approval for mine closures, and reduced mining company liability for clean-up and site remediation. Named in apparent seriousness, the Red Tape Reduction Bill of 1998 was even broader, amending 98 different statutes at once, including environmental measures. Arguably its most significant initiative was requiring provincial regulatory agencies to conduct cost-benefit analyses prior to setting any new administrative rule or standard. The measure obviously created obstacles to administrative rule setting. In combination with dramatic staff reductions, blistering ideologically motivated rhetorical attacks on public sector employees and their unions by Harris and his cabinet, and instructions given to provincial staff by senior bureaucrats and legislative staff not to prosecute violators of a range of environmental standards and rules (Krajnc 2000), the *laissez-faire* message was clear.[3]

The Harris administration also dismantled most of Ontario's environmental advisory boards and commissions, removing an important avenue for citizens to augment state capacities in social regulation. This includes termination of the Advisory Committee on Environmental Standards, the Environmental Assessment Advisory Committee, and the Ontario Roundtable on the Environment and the Economy (Krajnc 2000). The demise of these advisory boards and commissions – in combination with the elimination of provincial funding for citizen groups to organize and contribute to a range of provincial regulatory and administrative processes[4] – meant closure of critical avenues for independent scientific and public input, effectively consolidating the power of elite policy-makers within Harris's inner circle while freeing up private capital from the "burden" of state regulation.

Environmental governance reform under the Common Sense Revolution also included considerable zeal for outright privatization. These efforts include the so-called "Lands for Life" initiative, under which the province entered negotiations with forestry capital to create long-term tenures over vast provincial forest lands. Justified in terms of the Tragedy of the Commons discourse – a central discursive underpinning of environmental neoliberalism (Dryzek 1997) – the programme represented a marked departure from existing tenure rights, made all the more significant given that provincial forest lands cover fully half of Ontario.[5] Other fronts in the thrust toward privatization included the dismantling and sale of Ontario Hydro, previously one of the largest energy utilities in North America, as well as deepening earlier NDP gestures toward privatization of water and other utilities. Notably, the Omnibus Bill of 1996 repealed a Public Utilities Act provision requiring a public referendum prior to the sale of any utility, signalling the province's intention to make privatization easier. This was made more explicit by Bill 107 which fully divested the province of responsibility for sewage and water, and laid out guidelines for the outright sale of municipal utilities (since put on hold, largely because of Walkerton). Critically, privatization was justified based on largely ideological claims that public service provision is inherently inefficient, and specifically, anathema to the introduction of market mechanisms. Yet, it is critical to note that throughout this period, there was in place no provincial restriction on Ontario public utilities using marginal pricing schemes (C.N. Watson and Associates 2001).

All of these aspects of neoliberal reform provide the important context for the Walkerton tragedy. But the intersection of environmental neoliberalism in Ontario with the Walkerton tragedy is most immediate in the arenas of agricultural and municipal water regulation. Despite longstanding concerns about the problems posed by agricultural wastes in Ontario, and in the context of rising numbers of total livestock as well as farming intensity in the province (Beaulieu 2001; MacLachlan 2001; Miller 2000b; Winson 1993), very little was ever done prior to Walkerton by way of setting and enforcing binding standards on agricultural waste handling. In 1984, the Ontario Environmental Protection Act, which came to be known as the Ontario "spills bill," was passed, creating new standards and procedures on the storage and transport of hazardous wastes. But the government of the day bowed to Ontario's powerful farm lobby and exempted the farm sector.[6] As a result, livestock operations already enjoyed a comparatively *laissez-faire* regulatory milieu prior to the Common Sense Revolution. Nevertheless, the Harris administration's ideological zeal for private property rights and regulatory rollback further undermined the social regulation of agriculture. Fiscal and administrative downsizing crippled the OMAFRA, where the total staff was reduced to 661 positions by 2000, down two-thirds from a decade before. Moreover, backed by a still-potent farm lobby, in 1998 the Harris administration passed so-called "right-to-farm" legislation under the auspices of the Farming and Food Production Protection Act (FFPPA). The FFPPA (ironically, in an administration cloaked with the tropes of democratic devolution) centralized and bureaucratized the regulation of farm waste, blocking community-level complaints against farm operations, including those pertaining to manure handling and disposal, and created instead the Normal Farm Practices Protection Board at the provincial level (Miller 2000a). The Board, stacked by Harris administration appointees from the ranks of agro-industry, was given the authority to issue policy statements defining "normal" farm practices against which community complaints would be muted. This is one of the ways in which "*laissez-faire* was planned" (Polanyi 1944) in the agriculture sector, so that up to the time of the Walkerton tragedy, there were in fact no binding provincial requirements for manure storage or application (O'Connor 2002b).

In provincial regulation of municipal drinking water supplies, it was actually (as noted) the NDP that first allowed private water testing labs as an option to provincial labs. In addition, it was the NDP that devolved fiscal responsibility for water testing. But it was Harris's administration that then force-fed the market solution to municipalities by closing all three regional public water testing labs run by the MOE in 1996 – ending provincial testing altogether. Provincial intentions to completely privatize the utilities (see above) further entrenched the emerging model of devolved, and increasingly privatized service delivery in the water sector.

Critically, the Harris administration created what proved to be a lethal form of neoliberal reform combining privatization with *laissez-faire* deregulation. Despite forcing municipalities to contract out for water testing, the province passed no legislation or binding policy requiring either the municipality or the private labs to notify the province in cases of contaminated water. The province also failed to introduce any certification programmes for labs or their staff, nor provisions for inspection, nor any auditing procedures. In short, there was no oversight whatsoever. In fact, there were not even any legally binding water quality standards in place in May of 2000. Rather, despite a plethora of domestic and international evidence as to its potential toxicity, the MOE actually dropped E. coli from a provincial contaminants list under the Drinking Water Surveillance Program in 1996, just prior to the wholesale elimination of the programme altogether.[7]

An environment of risk and normal accidents

The sweeping climate of neoliberal environmental governance reforms imposed via the Common Sense Revolution provided both direct and proximate causation in making the Walkerton water poisoning a normal accident of regulatory failure. The fruits of these conditions for regulatory failure are evident in the Walkerton chronology.[8] On April 24, 2000, the town of Walkerton switched private water labs, from GAP Laboratories, an accredited water lab, to A&L Laboratories Canada East, a firm not accredited to conduct bacteria tests in Canada (but legally enabled to do so in Ontario). On May 5th, A&L found contamination in samples of treated Walkerton water, and notified the town PUC, but not the MOE nor the regional Medical Officer of Health. On May 16th, A&L Labs again notified the Walkerton PUC, citing counts in excess of 200 E. coli per 100 mL of treated water; again, the province was not informed. Two days later, A&L faxed the Walkerton PUC to communicate that the town's entire water system was contaminated. No notice was sent to the province. While it is true that during the crisis, information was intentionally obscured by PUC management negligence, and while Public Utilities Commission manager Stan Koebel did lie about the results of water tests, at no time was there a regulatory requirement to notify the province. Moreover, there was no system in place for evaluating and weeding out people like the Koebel brothers who were categorically unqualified for staffing a municipal water utility (O'Connor 2002a).

The absence of binding standards for agricultural waste disposal and the lack of a groundwater protection plan, particularly for karst areas, are also clearly pertinent issues of regulatory context. As part of the Walkerton investigation, DNA tests matched bacteria recovered from Walkerton's water with cattle from a single farm next to Well 5, a farm owned and operated by local resident Dr. David Biesenthal. The farm in question was not a particularly large one, featuring a herd of about 95 animals on 54 ha (O'Connor 2002a), hardly an industrial operation. Moreover, the owner of the farm appears to have followed provincial guidelines on agricultural waste disposal (O'Connor 2002a). Yet, if a relatively small farm producing moderate amounts of waste according to provincial guidelines could nevertheless so severely contaminate a town's drinking water, this should only underline the potential risks associated with larger operations and volumes of waste, and the extreme vulnerability of this particular hydrological regime.

Conclusion

Inevitably, I have omitted from this narrative essential elements of the Walkerton story, and of environmental neoliberalization in Ontario. This includes important political and regulatory responses to Walkerton's tragedy. Subsequent to a high profile provincial inquiry into the Walkerton affair, Inquiry Commissioner Dennis O'Connor (2002a; 2002b) recommended specific changes in provincial regulations, notably helping to precipitate the Nutrient Management Act of 2002. This legislation committed the province to binding standards for manure spreading and disposal, featuring setbacks from surface water for manure spreading and caps on the total amounts of manure that can be spread per unit area of land. In addition, the capacity and courage shown by many residents of the Walkerton area – some of whom attempted to redress regulatory gaps discussed here prior to the tragedy – have been inspiring, particularly as led by the Concerned Walkerton Citizens coalition. Although I have not focussed on citizen mobilization, I do not mean to suggest its absence.

What I have tried to emphasize and demonstrate here is that the scope and character of environmental re-regulation under the Common Sense Revolution made Walkerton a normal accident of neoliberalism, in turn reinforcing the observation that socio-natural relations and environmental governance are central (i.e. not incidental) projects in the neoliberal turn (McCarthy and Prudham 2004). Despite clear evidence that local factors were central in shaping how, where, and why the tragedy occurred, the broader context of environmental governance, and the organized irresponsibility introduced under the Common Sense Revolution were also clearly implicated. Pertinent issues of regulatory context include a chill placed on the establishment and enforcement of environmental regulations. Moreover, gaps in regulatory oversight were in significant measure opened or widened by neoliberal reforms, including in the areas of agricultural waste disposal, groundwater management, and in provincial oversight vis-à-vis municipal drinking water utilities.

Stepping back slightly from the immediacy of Ontario's experience, and returning to Beck's notion of the Risk Society, it may be that environmental risks indeed do comprise one of the central themes in the politics of late capitalist modernity. If so, however, it may also be true that these risks and these politics are less dissociated from capitalism and its social regulation, particularly in an era of neoliberal reform, than Beck would have us believe.

All of this serves to reinforce some themes that, sadly, should have been known prior to Walkerton. *Laissez-faire* faith is simply not enough to enable nominally free markets alone to act as a form of social regulation, contrary to the self-evident logic suggested by the "Common Sense" Revolution. And while *laissez-faire* ideas continue to hold powerful political and discursive appeal, as the Ontario experience clearly demonstrated, the prescriptions of *laissez-faire* can be exposed as contradictory, irresponsible, and highly politically interested when analysed in specific social and ecological contexts; this is one of the aims of this chapter, and of this volume. If any good can come from an incident such as the one that occurred in Walkerton in May of 2000, it may be to serve as a reminder that state-centred social regulation of the environment, at a minimum, ought to be oriented toward protecting society against the excesses of what Polanyi called market self-regulation, rather than the other way around.

Acknowledgements

I would like to thank Bruce Braun, James Glassman, Andrew Leyshon, James McCarthy, Juanita Sundberg, and two anonymous referees for providing comments and suggestions on earlier drafts of this chapter. Any remaining errors or ambiguities are my responsibility alone. A longer version of this chapter was Prudham, S. (2004) "Poisoning the Well: Neoliberalism and the Contamination of Municipal Water in Walkerton, Ontario," *Geoforum* 35(3): 343–59.

Notes

1 Some have permanent organ damage, particularly to their kidneys. Others cite persistent after-effects from the trauma of the experience, including paranoia regarding every little illness, and an acute, lasting distrust of drinking water. One resident stated before the public inquiry into the tragedy, "Oh my goodness, why am I so depressed and how come I can't stop crying? It's scary just not knowing what's going to happen ... next" (O'Connor 2002a: 44).

2 This individual strain of E. coli has been known to be particularly virulent since it was first identified in an outbreak of E. coli contamination linked to beef consumption in Oregon in 1984.

3 Canadian Environmental Law Association 1999, "Guide to Environmental Deregulation in Ontario", available at: http://www.cela.ca/appendix.htm. Accessed November 9, 2000.

4 Accomplished by allowing the Intervener Funding Act to expire.

5 www.cela.ca/appendix.htm, op. cit.
6 Critical to the politics of this exemption was the farm lobby's familiar discursive representation of farmers as a yeoman class of small, independent producers in need of protection from the "burdens" of environmental regulations, a characterization clearly problematic in the context of increasingly industrial, and in some instances, thoroughly corporate farm operations.
7 Ulli Diemer, "Contamination: the Poisonous Legacy of Ontario's Environmental Cutbacks", *Canadian Dimension*, July/August 2000, 34(4): 33–5.
8 Based on the inquiry report (O'Connor 2002a) and "The Walkerton Story", *Toronto Star*, Saturday October 14, 2000.

References

Beaulieu, M. S. (2001) *Intensive Livestock Farming: Does Farm Size Really Matter?* Ottawa: Publisher Statistics Canada, Agriculture Division Working Paper No. 48.

Beck, U. (1999) *World Risk Society*, Cambridge: Polity Press.

Beck, U. and Ritter, M. (1992) *Risk Society: Towards a New Modernity*, London: Sage Publications.

Benton, T. (1997) "Beyond Left and Right: Ecological Politics, Capitalism and Modernity," in M. Jacobs (ed.) *Greening the Millennium: The New Politics of the Environment*, London: The Political Quarterly Publishing Co., pp. 34–46.

C.N. Watson and Associates (2001) *Financial Management of Municipal Water Systems in Ontario*, Mississauga, Ontario: Prepared for the Canadian Environmental Law Association

Cowell, D. W. and Ford, D. C. (1980) "Hydrochemistry of a Dolomite Karst – the Bruce Peninsula of Ontario," *Canadian Journal of Earth Sciences* 17(4): 520–6.

—— (1983) "Karst Hydrology of the Bruce Peninsula, Ontario, Canada," *Journal of Hydrology* 61(1–3): 163–8.

Drew, D., Hötzl, H. and International Association of Hydrogeologists (1999) *Karst Hydrogeology and Human Activities: Impacts, Consequences and Implications*, Brookfield, VT: A. A. Balkema.

Dryzek, J. S. (1997) *The Politics of the Earth: Environmental Discourses*, Oxford: Oxford University Press.

Ford, D. and Williams, P. W. (1989) *Karst Geomorphology and Hydrology*, Boston, MA: Unwin Hyman.

Jessop, B., Bonnett, K. and Bromley, S. (1990) "Farewell to Thatcherism? Neo-Liberalism and New Times," *New Left Review* 179: 81–202.

Keil, R. (2002) "'Common-Sense' Neoliberalism: Progressive Conservative Urbanism in Toronto, Canada," *Antipode* 34(3): 578–601.

Krajnc, A. (2000) "Whither Ontario's Environment? Neo-Conservatism and the Decline of the Environment Ministry," *Canadian Public Policy* 26(1): 111–27.

McCarthy, J. and Prudham, S. (2004) "Neoliberal Nature and the Nature of Neoliberalism," *Geoforum* 35(3): 275–83

MacLachlan, I. (2001) *Kill and Chill: Restructuring Canada's Beef Commodity Chain*, Toronto: University of Toronto Press.

Miller, G. (2000a) *Changing Perspectives: Environmental Commissioner of Ontario Annual Report 1999/2000*, Toronto: Government of Ontario.

—— (2000b) *The Protection of Ontario's Groundwater and Intensive Farming*, Toronto: Environmental Commissioner of Ontario, Government of Ontario.

O'Connor, D. R. (2002a) *Report of the Walkerton Inquiry: The Events of May 2000 and Related Issues*, vol. 1, Toronto: Ontario Ministry of the Attorney General.

—— (2002b) *Report of the Walkerton Inquiry: A Strategy for Safe Drinking Water*, vol. 2, Toronto: Ontario Ministry of the Attorney General.

Peck, J. (2001) "Neoliberalizing States: Thin Policies/Hard Outcomes," *Progress in Human Geography* 25(3): 445–55.

Peck, J. and Tickell, A. (1992) "Local Modes of Social Regulation? Regulation Theory, Thatcherism, and Uneven Development," *Geoforum* 23: 347–64.

—— (1995) "The Social Regulation of Uneven Development: 'Regulatory Deficit', England's South East, and the Collapse of Thatcherism," *Environment and Planning A* 27: 15–40.

—— (2002) "Neoliberalizing Space," *Antipode* 34(3): 380–404.

Perrow, C. (1999) *Normal Accidents: Living with High-Risk Technologies*, Princeton, NJ: Princeton University Press.

Polanyi, K. (1944) *The Great Transformation: The Political and Economic Origins of Our Time*, Boston, MA: Beacon Press.

Smith, N. (1984) *Uneven Development: Nature, Capital, and the Production of Space*, Oxford: Basil Blackwell.

Tickell, A. and Peck, J. (1995) "Social Regulation after Fordism: Regulation Theory, Neo-Liberalism, and the Global-Local Nexus," *Economy and Society* 24(3): 357–86.

Winson, A. (1993) *The Intimate Commodity: Food and the Development of the Agro-Industrial Complex in Canada*, Toronto: Garamond Press.

Worthington, S. R. H., Smart, C. C. and Ruland, W. (2001) *Karst Hydrogeological Investigations at Walkerton*, Walkerton, Ontario: Concerned Walkerton Citizens.

14 Un-real estate

Proprietary space and public gardening

Nick Blomley

Introduction

Contemporary urban governance and policing strategies, with their embrace of public–private partnerships, deregulation, fiscal austerity, cross-subsidies, and market solutions have been characterized as a form of urban neoliberalism (Peck *et al.* 2001). Neoliberalism is, in part, a language of property – a return to central axioms of eighteenth-century liberalism, which locates private property as the foundation for individual self-interest and optimal social good (Smith 2002: 429). Similarly, much contemporary urban policy turns on property, whether it is enacted in order to protect private property, sustained according to the logic of property, or entailing the extension of private property to public domains (Rogerson and Boyle 2000). However, geographers have had little to say about property in general, or considered the ways in which it has been put to work in the contemporary city.

Although Crime Prevention Through Environmental Design (CPTED) predates neoliberalism, its principles have been taken up in contemporary public order regulation and policing. This chapter seeks to demonstrate the ways in which CPTED relies upon particular geographies of property – as individualized and certain. This theorization of property, I argue, fails to acknowledge the diverse, contradictory and sometimes collectivized ways in which property can be put to work. I demonstrate its limitations through a case study concerning an attempt to enrol urban residents in a public gardening project, termed a greenway. Property, at the core of the neoliberal urban agenda, is a more diverse category, both empirically and politically, than we often suppose (Blomley 2004b). Therein, perhaps, lies its radical potential.

CPTED, defensible space and private property

In 1995, some residents on a street in a low-income, inner city neighbourhood of Vancouver, Canada, called Strathcona, approached the City under its Neighbourhood Greenway project. The programme allows for partial funding and design, development and construction assistance from the City, with residents providing funds and/or in-kind contributions to planning, construction and maintenance of the greenway.[1] After some controversy, a greenway was created in 1997. As one planner put it, the greenway was supposed to demonstrate the truth of what he termed "'the broken window theory': actually having people out there caring for it sends a very strong message that this is territory that is being cared for, it's not open game" (interview, July 30 2000). At a public meeting, convened following the concern of some residents that the greenway would attract "undesirables," a City Police Officer spoke on the merits of these forms of "ownership." Unlike more passive forms of community securitization, such as community policing, the act of gardening was seen as a means by which these enactments would occur.

The language here echoes an assumed relationship between property, public space and disorder, which has been popularized by the Crime Prevention Through Environmental Design (CPTED) movement. CPTED derives from the work of writers such as Jane Jacobs, who wrote of the "marvellous ... complex order" (1961: 50) that encourages "eyes upon the street, eyes belonging to those we might call the natural proprietors of the street" (ibid.: 35). The principles of CPTED have been applied in various ways, and under different rubrics; however, a strong case can be made that Oscar Newman's principles of defensible space are at its core.

Broadly defined, defensible space is said to concern a set of territorial mechanisms – real and symbolic barriers, "areas of influence," and forms of surveillance – that "combine to bring an environment under the control of its residents" (Newman 1972: 3). Design elements, such as the positioning of windows, allow residents to survey public space. However, central to Newman's treatment of territory is the argument that design must encourage residents to extend the property claim they have to their private space to adjacent public space. This reflects the maxim, as one CPTED primer puts it, that "people will take care of space and assets in which they have a proprietary concern" (Crowe 1991: 103). This extension is said to deter potential offenders: "The better a place is defined regarding ownership, the more likely a non-resident or visitor is to be conspicuous" (Crowe 1991: 104). Unlike the actual privatization of public space noted in other settings, this extension is provisional and extra-legal. Residents' claim to adjacent space is, in that sense, a form of "illusory property," or "un-real estate."[2]

The territorialization of property relations, moreover, appears to be pre-social. One CPTED primer argues that humans have "a need to establish both temporary and permanent ownership of space ... Humans and animals mark their turf" (Crowe 1991: 99). Design, in this sense, is not manipulative, but rather "catalyze[s] the *natural* impulses of residents" (Newman 1972: 11, my emphasis). Yet while "ethnic and cultural divisions provided previous generations of city residents with a form of solidarity" (1973: 13), territorially defined space has been eroded in the modern city,

Newman argues. "In our newly-created dense and anonymous residential environments, we may be raising generations of young people who are totally lacking in any experience of individuality, of personal space, and by extension, of the personal rights and property of others" (1972: 4). What is needed, then, is to reintroduce principles of territorial definition that reflect our human (and perhaps biological) legacy.

It is clear that the defensible space is understood as exclusively individualized and quasi-private, reliant upon a particular set of spatial arrangements that "*extends the area of the residential unit into the street* and within the area of felt responsibility of the dweller" (1972: 4, my emphasis). Of course, this extension into public settings generates an interesting tension, given that public space is conventionally assumed to be a shared and collective resource. However, Newman still insists on treating it in essentially individualized terms. Design principles, he argues, in one particularly convoluted moment, should make public space "the shared extension of the private realms of a group of individuals" (1972: 2). He is clearly uncomfortable with the possibility of a faceless collective, suggesting that there is "an upper limit, an entropy principle, beyond which the critical mass becomes a collection of homogenous individuals who bear no relationship to one another, and who do not participate in a sense of collective responsibility" (1973: 17).

Indeed, Newman seems unwilling to accept the possibility of any *collective* entitlement to space. "Proprietary attitudes" are imagined as self-regarding; they concern the owner and that which he owns, set against those who might threaten those entitlements. Conversely, as property theorist Carol Rose notes, the commons is seen as a space of "violence and danger" (1994: 291). Newman's approach reflects the pervasive assumption that while there may be many owners of land, there are "for practical purposes, ... only two classes of ownership" (Geisler 2000: 65). On the conventional account, "markets are based on private rights, or, when markets fail, property may be governmentally managed" (Rose 1994: 110). Other possibilities (such as non-state common property) are deemed vestigial or even incoherent.

Newman notes in particular the significance of garden space, and the act of gardening as a means by which this process of privatized extension can occur. In his treatment of the private home, the preeminent statement of territorial claim, he suggests that the garden, acts as a "penumbra of safety" (1973: 16) "reinforced by symbolic shrubs or fences" (1972: 51) that creates an important buffer with others. Unfortunately, in high-density urban settings, or public housing, this periphery is absent. In these environments he advocates the creation of buffers, deploying gardening, on occasion, as a means by which a proprietary claim can be enacted. Property, in this sense, is not a static statement, but requires active forms of human doing that constitutes the self at the same time as it communicates a claim to others (Rose 1994).

Territorial definitions, if they are to effect behaviour, must be certain and non-ambiguous. Newman stresses the importance of legible symbols which define clear spatial boundaries, marking transitions from public to private spaces: They serve to "inform that one is passing from a space which is public where one's presence is not questioned through a barrier to a space which is private and where one's presence requires justification" (1972: 63). By definition, space that is "unassigned" is problematic. Gardening, with its evidence of "improvement" and its marked boundaries, is one useful means of communicating these designations, definitions and designs to the extent that it signals a legible and clear statement of ownership (Blomley 2004a).

In all cases, there is an assumption of spatial legibility and clarity to property and its boundaries. Property, whether public or private, must be clearly assigned and certain. This is a familiar legal world of bright lines and determinacy. The assumption that property can be clearly delineated into private and public realms, of course, is pivotal to liberalism. The supposed definitional certainties of property – Bentham's "established expectations" (1843/1978: 52) – are valuable to social ordering. The security of knowing which is mine, and which is thine is said to reduce the possibility of conflict.

The three principles – that property is largely synonymous with private property, that it is communicated to others through clear acts, such as gardening, and that it is, or should be, definitionally certain – accord with hegemonic accounts of property, characterized usefully by Singer (2000) as the ownership model. This sees property premised on consolidated, permanent rights vested in a single identifiable owner, identified by formal title, exercising absolute control, distinguished from others by boundaries that protect the owner from non-owners by granting the owner the power to exclude. The owner is also marked off from all others: the actions of the owner are self-regarding: they concern him or herself alone and the things owned. Property, in that sense, is almost exclusively *private* property. The centrality of the ownership model renders other modalities of ownership invisible. Common property, declares one scholar, "means no property" (Harris 1995: 438). The ownership model presents property as fixed, natural and objective, transforming the historical and social contingency of social history into limited structural arrangements and ideological commitments (Blomley 2004b).

The Atlantic Street Greenway

Gardening presents a useful and accessible category for exploring the ways people think about and act in relation to property, in part because of its practical, embodied and geographical qualities. As I have noted above, gardening is also touted as a means of enacting a proprietary claim. My assessment of the Atlantic Street Greenway in Vancouver is part of a larger research project conducted in Strathcona, a largely low-income, ethnically diverse neighbourhood marked by beginnings of white and middle-class gentrification. Drawing from interviews and participant observation through my involvement as a resident of Atlantic Street, the following section investigates accounts from residents and civic officials regarding notions of public and private space, property and entitlement in relation to the greenway.

From a CPTED perspective, the domestication of a "wild" space, and its conversion into a "gardened" space, over which local residents would have a claim, by virtue of their "investment" in the greenway were intended to signal a property claim that would persuade the disorderly to go elsewhere. Superficially, there is some apparent evidence for the success of the greenway on these terms. In a public meeting one resident noted that the greenway

> would enable us to claim ownership of this strip of neglected land and send a message to [drug-users, prostitutes, etc.] that this is a place where people "*care*" ... With residents maintaining the Greenway the result will bring ... a responsibility for the street as a whole rather than for only your "own" property.
>
> (City of Vancouver 1997: no page, original emphasis)

Viewing the greenway retrospectively, Don, a resident, seemed to agree:

> I think it's really improved the look of the whole block and I think it has done what we hoped it would do, which would be kind of send out a message that this is a cared for area, and it's not derelict and it isn't going to go unnoticed if somebody wants to do something criminal there.

Nearly all of the residents interviewed felt that the greenway had been an improvement. Previously, as one put it, the site had looked "untended and anonymous." Now the area looked "more cohesive": "it just looks more of a residential ... sort of a nicer setting."

True to the individualized logic of defensible space, the original impetus for the greenway came from one resident who had taken it upon himself to scythe the grass and clean up rubbish. However, although the catalyst for the greenway was an individual, engaged in cultivation, Newman's notion of the private claiming of space did not seem realized. Many respondents were explicit in arguing that their relationship was not individualized:

Q: Do you feel like it's kind of "yours" now?
DARLENE: It's definitely a community thing, but I don't feel like I "own" it, I feel like we share it.
Q: Does it feel like it's "your" space?
JOHN: Not mine personally, but it feels like it belongs to everyone on the street. It's ours.

The exclusionary mindset imagined by Newman is also not in evidence. While for some, as noted, a space that appeared "cared for" was less likely to be "abused," others worried at the notion that the greenway would repel the "anti-social." For them, the goal was to create a space that could be attractive, but not necessarily exclusive.

CAROLYN: Antisocial activity ... I'm a bit torn about that.
Q: In what way?
CAROLYN: Well, I look at, for instance, how they clear by the viaduct [a nearby road, cleared of brush] so nobody can sleep out there. People who have nowhere to sleep ... Why not let them have somewhere, at least in summer, to sleep out under?

The act of working the land had, indeed, vested title in those who had transformed it. Contra CPTED, however, this was a collective and non-exclusive entitlement that overlay formal public ownership in a complex set of overlapping estates.

DON: [I]t's common land, it belongs to everyone. Legally it belongs to the city of Vancouver, but I think people think of it as belonging to the neighbourhood because the neighbourhood created it, and worked quite hard to build it.
Q: So you think that once the residents have worked on it, it makes it theirs?
CAROLYN: Well, yes, they've earned the right to enjoy the land. They are improving it, they are beautifying it, they're doing the city's work for them.

This entitlement, by extension, was only as good as the continued enactments that made it "ours":

Q: Whose land is it?
JOHN: It doesn't matter whose it is – we're doing stuff there, it's ours for now. Yet these collective claims were also cross-cut by more individualized entitlements, based on particular patterns of use and investment. An exchange between Flora and John indicated a sense of ownership enacted by individual acts of appropriation, such as Margaret's "garden" or John's apple tree. However, even here it was hard to identify the owner: Margaret's curly willows were also partly John's, he joked. A pear tree was public, but the pears were John's, though he was happy to share. If the greenway is "illusory property," in other words, it is owned in multiple and intersecting ways

But the question of who "owns" the street is even more complex. As far as the City is concerned, local ownership is provisional and partial. While one planner saw the greenway and related initiatives as pushing the envelope, a "letting go of fear," there is still a sense that the private and public are to be clearly delimited and separate, with the public realm held as "municipal property" (interview, July 30 2000). As Macpherson (1978) notes, state ownership in this sense is of a particular form, entailing a right that the state creates and retains for itself (for example, public utilities, or parkland). In effect, the state is acting as an artificial person, or corporation. This makes it possible for the City to claim a tree planted in public as "our tree," as if it had been planted on private property:

> People have often planted trees behind the sidewalk kind of thing, that have become public trees, sometimes to their sorrow because then they'll want to cut it down and we'll say you can't cut it down because it's our tree. It belongs to the public now and we've maintained it, we're going to maintain it and the fact that you planted it, *it's no different than if you planted it on your neighbour's property.* If you wanted it to be under your control you should have put it on your property.
> (Interview, Parks Board official, July 28 2000, my emphasis)

This clear separation between private and municipal property also seemed to be a priority for local opponents of the greenway. As noted, there was far from unanimous support for the creation of the greenway amongst local residents. It is hard to establish the logic – in part, perhaps, there was a sense that newer residents were unduly throwing their weight around. However, this also seemed underlain by the claim that private and public needed to be clearly distinguished, and that acts such as community gardening or the greenway were illegitimate private takings of a pure public good. For some, only the rightful owners, the state, should conduct public gardening. When asked about private individuals planting in public, Ray, a Chinese-Canadian senior who lived on Atlantic Street was clear:

> I don't think it's right. It should simply be grass. Just green. This is privately grown. If it is on government land, then it is not right at all. It should be grass. Then the entire city would be green. The government should lay grass there.[3]

Property was to be exclusively confined to private space: indeed, some worried that their enjoyment of their private property would be compromised by the greenway, which would attract transients and other threatening types.

While my analysis is based on a small sample, and is thus not necessarily generalizable, I trust it begins to suggest some empirical complexities that depart from Newman's treatment of defensible space. While there is a sense of "ownership" on the Atlantic Street Greenway, then, it is far from straightforward. It does not appear individualized; or rather, if it is, this is overlain by collective understandings. The greenway "belongs" to the municipality, the public, the neighbourhood, the street, some (largely

Anglo) residents, John, private owners and whoever felt "the most nurturing towards one particular area." The boundaries of property are uncertain, intersubjective, and layered. Although initiated by one individual, public gardening created overlapping and multiple claims that were both individual and collective.

Alternative forms of public gardening

The language of broken windows, CPTED and defensible space has become so prevalent as to appear commonsensical. As I have suggested here, this may be because of its roots in certain taken-for-granted notions of space and property. As a consequence, it becomes hard to imagine alternative analyses. However, Newman's analysis and related neoliberal initiatives can productively be contrasted to other scholarly attempts to address greenways and public safety. These re-conceive the relation between territory, property and the public domain. What is particularly interesting for my purposes is the ways in which property is also treated in a more diversified and socialized way.

One interesting Canadian example is that of the Safer Cities Movement that, according to its proponents, offers an alternative to conventional approaches to law and order, given its attention to marginalized groups, including women and the elderly, its reliance upon a mix of both social and physical interventions, and its orientation to community mobilization. While Safer Cities' approaches may incorporate some CPTED principles, the emphasis may differ. Wekerle and Whitzman (1995) acknowledge the potential benefits of territoriality as a form of crime prevention, but caution that enclosed spaces may actually increase vulnerability to crime, and view territory as necessarily collective, open, and formed through inclusive policy.

Jane Jacobs' conception of "eyes on the street," claimed by Newman and others as a progenitor of CPTED's notion of territoriality, is also reworked in this literature, it being argued that CPTED ignores the subtlety of Jacobs' argument, and her injunction against the claiming of "turf" (Wekerle 2000a). There have also been attempts to re-conceptualize greenways in the Canadian context through the Safer Cities prism: while attentive to design and environmental conditions, the argument here focuses on the differentially gendered nature of urban space, as well as the importance of community mobilization and input. Although there is an attempt to achieve local ownership and stewardship, this is not individualized:

> In terms of actual and perceived safety, the most important stakeholders in urban greenways are the users themselves. Those users who are most sensitive and vulnerable in society – women, children, those with physical disabilities and aging people – are often the most reliable barometers of greenway safety.
>
> (Luymes and Tamminga 1995: 396)

The central issue, it is argued, is "personal control and freedom of use, essential to the development of 'democratic' places in increasingly diverse urban communities" (Luymes and Tamminga 1995: 400, cf. Wekerle 2000b). Carol Rose (1994) reflects on the work of William Whyte who, not coincidentally, coined the phrase "greenway" (Little 1990: 23). Whyte considers the way design can be used to invite urbanites to "make themselves at home" in public places in ways that, while offering some level of protection and security, encode spatial invitations to participate and co-inhabit.

Such prescriptions, ironically, are not too dissimilar from the original conception of the greenway within Vancouver. Moura Quayle, a University of British Columbia Landscape Architect, chaired the Urban Landscape Task Force that first proposed greenways (termed "public ways" in the report to signal their link to her vision of an enriched public realm). She is openly hostile to CPTED, which she sees as a sterilization of space that places blame on the landscape (interview, 14 September 2000). She cautions against seeing the public domain as a fearsome and incomprehensible zone, and encourages a view of greenways as productive of citizenship and community. Greenways and public ways are "the hearts and minds of people – an urban attitude characterized by cohesion, pride, identity and community life" (Quayle 1995: 468) and the rebuilding of citizenship. The Task Force report uses a similarly collective language, with calls to create "truly public spaces" and "democratic streets." Public space is not simply "municipal property" but is imagined as in the "public realm," spaces of shared citizenship and the being together of strangers. Private property rights must accommodate the needs of the collective: for example, it argues that property owners who clear their land of trees do ecological and visual damage to the larger community.

In this, the Task Force echoes some of the language of the U.S. greenway movement. While much of the literature on greenway movement acknowledges the realities of private property, encouraging greenway activists to negotiate creatively with private property owners, Little also argues that the greenway should utilize what he terms the "linear commons of a community" (1990: 33), such as riparian lands and abandoned right-of-ways, over which some form of public interest is recognized in law and policy as well as linear routes over which the public has a less formalized, but still viable claim, such as land along a road that

> is agreed to be historic or scenic [and] also has common value, as part of the public landscape, in maintaining a community's sense of itself. Although the title to ownership of such lands may be lodged in private hands, the public's interest in their use and conservation is generally understood.
>
> (1990: 34)

These alternative accounts make a connection between space and property, but depart from the reliance upon the individualized propriety spaces of mainstream CPTED. Property is not reduced to the state–private binary. Instead, we find other possibilities, including the recognition of forms of ownership that is neither. Such claims are not unprecedented: the public trust doctrine within Anglo-American property law vests ownership in neither the state nor private hands, but in the public (Sax, 1970). Ownership need not only be vested in private parties or organized governments: " ... these two options do not logically exhaust all the possible solutions ... the common law ... with surprising consistency, recognizes two distinguishable types of *public* property" (Rose 1994: 110).

One can discern, in other words, a form of ownership that exists "somewhere between individual private property on the one hand and the tragic commons on the other" (Rose 1994: 292; 1998). Such arrangements, it has been suggested, are ubiquitous, though often overlooked (Ingerson 1997). "The commons," it has been argued, " ... is an underrated, much-ignored reservoir of valuable resources, system of social governance, and crucible for democratic aspirations" (Steinberg 1995: 15). Such claims, of course, sounds hopefully old-fashioned and fragile, especially given the supposed death of public space, lamented by many. And there is much to lament when, for example, business associations privatize parks, or police show "zero tolerance" for the homeless. But this supposes that the commons is a static space, rather than something produced through social action, such as public gardening. It also characterizes public land as municipally owned land, such as the street. However, as Atlantic Street shows, "private owners" can convert municipal property into a commons. And as I have tried to demonstrate elsewhere, private spaces can also be laid claim to in the name of a local community (Blomley 2004a) while "private" owners can think of "their" land in relational and socialized ways (Blomley 2004b, 2005a, 2005b).

Conclusion: the diverse politics of property

Neoliberalism is not simply a set of "economic" beliefs and practices, such as a faith in the market as an ordering mechanism. It also draws from political and legal understandings, of which property is perhaps one of the more important. Property, as a set of practices, rhetorical claims and legal relations can – of course – be used for socially harmful ends, as has been demonstrated by a number of geographers (Mitchell 2003; Blomley 2003b). To characterize property *only* in these ways, however, is to accept a neoliberal definition. It is akin to treating categories such as "citizenship," "rights," or "economy" according to hegemonic prescriptions. The danger, of course, is that we lose sight of the diversity of such categories. More productive is Gibson-Graham's suggestion (1999: ix) that we "depict social existence at loose ends with itself, rather than producing social representations in which everything is part of the same complex and therefore ultimately 'means the same thing' (e.g. capitalist hegemony)" Refusing such representations, Gibson-Graham (1998: xi) suggests, allows us to "represent economic practice as comprising a rich diversity of capitalist and non-capitalist activities that had until now been relatively 'invisible' because the concepts and discourses that could make them 'visible' have themselves been marginalized and suppressed."

In a recent editorial, Wendy Larner (2003) worries that recent research has treated neoliberalism as monolithic and unstoppable. She urges critical scholars to attend to the diversity, hybridity and contradictions of neoliberalism. As shown here, it can also usefully concern governmental programmes themselves, exploring the messy ways in which (largely) private property owners are enrolled in an attempt at regulating public disorder. When spatialized, property appears a more protean category. The spaces and subjects of neoliberalism, in other words, may be more slippery than we think.

The politics of neoliberalism, especially in relation to property, should not be taken as a settled fact, but as an analytic question. Let us not take them at face value. Even neoliberalism can contain, if you like, forms of neo-socialism. Crime Prevention Through Environmental Design can mutate into Collective Property Through Environmental Design. The power of prevailing categories of property is such that these alternative possibilities – that are both external to "private property" and also internal – are easily rendered invisible. If we are to give capitalism an identity crisis (Gibson-Graham 1998: 260), we will need to unsettle some core political categories, like property, and acknowledge their diversity and loose ends.

Acknowledgements

A longer version of this chapter was Blomley, N. K. (2004) "Un-real estate: Proprietary Space and Public Gardening," *Antipode* 36(4): 614–41. I am extremely grateful for the assistance of Oscar Newman, Gerda Wekerle, Lorraine Gibson, Moura Quayle, Alan Duncan, Pat Brantingham, Albert Klassen and three anonymous reviewers. Thanks also for the research assistance of Lorraine Gibson, Milo Wu, Aurian Haller and Kathleen Yan, and the generosity of those Strathcona residents who participated in this study. Research was funded by the Social Science and Humanities Research Council. The original version of this chapter provides considerably more detail on property, space and CPTED.

Notes

1 www.city.vancouver.bc.ca/engsvcs/streets/greenways/.
2 The characterization is Rose's (1994: 290).
3 Translated from the original Cantonese.

References

Bentham, J. ([1843], 1978) "Security and Equality of Property," in C. B. Macpherson (ed.) *Property: Mainstream and Critical Positions*, Toronto: University of Toronto Press.

188   *Nick Blomley*

Blomley, N. (2003) "Law, Property and the Geography of Violence: The Frontier, the Survey and the Grid," *Annals of the Association of American Geographers* 93(1): 121–41.
—— (2004a) "The Boundaries of Property: Lessons from Beatrix Potter," *The Canadian Geographer* 48(2): 91–100
—— (2004b) *Unsettling the City: Urban Land and the Politics of Property*, New York: Routledge.
—— (2005a) "The Borrowed View: Privacy, Propriety, and the Entanglements of Property," *Law and Social Inquiry* 30(4): 613–63.
—— (2005b) "Flowers in the Bathtub: Boundary Crossings at the Public–Private Divide," *Geoforum* 36(3): 281–96.
City of Vancouver (1997) *Administrative Report, to Standing Committee on Planning and Environment from General Manager of Engineering Services and Director of Community Planning*, September 24, Vancouver: City of Vancouver.
Crowe, T. D. (1991) *Crime Prevention Through Environmental Design*, Boston, MA: Butterworth-Heinemann.
Geisler, C. (2000) "Property Pluralism," in C. Geisler and G. Daneker (eds) *Property and Values: Alternatives to Public and Private Ownership*, Washington, DC, Island Press, pp. 65–86.
Gibson-Graham, J. K. (1998) *The End of Capitalism (As We Know it): A Feminist Critique of Political Economy.* Oxford: Blackwell.
Harris, J. W. (1995) "Private and Non-Private Property: What is the Difference?" *Law Quarterly Review* 111: 421–44.
Ingerson, A. (1997) "Urban Land as Common Property," originally published in *Land Lines*, 9, (www.lincolninst.edu/landline.1997/march/commonp.html), accessed June 6 2002.
Jacobs, J. (1961) *The Death and Life of Great American Cities*, New York: Vintage.
Larner, W. (2003) "Neoliberalism?" *Environment and Planning, D, Society and Space* 21: 509–12.
Little, C. (1990) *Greenways for America*, Baltimore, MD: Johns Hopkins University Press.
Luymes, D. T. and Tamminga, K. (1995) "Integrating Public Safety and Use into Planning Urban Greenways," in J. G. Y. Fabos and J. Ahern (eds) *Greenways: The Beginnings of an International Movement,* Amsterdam: Elsevier, pp. 391–400.
Macpherson, C. B. (ed.) *Property: Mainstream and Critical Positions*, Toronto: University of Toronto Press.
Marx, K. (2000) *Karl Marx: Selected Writings*, ed. D. McLellan, Oxford: Oxford University Press.
Mitchell, D. (2003) *The Right to the City: Social Justice and the Fight for Public Space*, New York: Guilford Press.
Newman, O. (1972) *Defensible Space: Crime Prevention Through Urban Design*, New York: Macmillan
—— (1973) *Architectural Design for Crime Prevention*, Washington, DC: U.S. Department of Justice.
—— (1995) "Defensible Space: A New Physical Planning Tool for Urban Revitalization," *Journal of the American Planning Association* 61(2): 149–55.
—— (1996) *Creating Defensible Space*, Washington, DC: U.S. Department of Housing and Urban Development.
Peck, J. (2001) "Neoliberalizing States: Thin Policies/Hard Outcomes," *Progress in Human Geography* 25(3): 445–55.

Quayle, M. (1995) "Urban Greenways and Public Ways: Realizing Public Ideas in a Fragmented World," in J. G. Y. Fabos and J. Ahern (eds) *Greenways: The Beginnings of an International Movement*, Amsterdam: Elsevier, pp. 461–75.

Rogerson, R. and Boyle, M. (2000) "Property, Politics and the Neo-Liberal Revolution in Urban Scotland," *Progress in Planning* 54: 133–96.

Rose, C. (1994) *Property and Persuasion: Essays on the History, Theory and Rhetoric of Ownership*, Boulder, CO: Westview Press.

—— (1998) "The Several Futures of Property: Of Cyberspace and Folk Tales, Emission Trades and Ecosystems," *Minnesota Law Review*, 83(129): 129–82.

Sax, J. L. (1970) "The Public Trust Doctrine in Natural Resource Law: Effective Judicial Intervention," *Michigan Law Review* 68: 471–566.

Singer, J. (2000) *Entitlement: The Paradoxes of Property*, New Haven, CT: Yale University Press.

Smith, N. (2002) "New Globalism, New Urbanism: Gentrification as Global Urban Strategy," *Antipode* 34(3): 452–72.

Steinberg, (1995) *Slide Mountain or the Folly of Owning Nature*, Berkeley, CA: University of California Press.

Waldron, J. (1988) *The Right to Private Property*, Oxford: Oxford University Press.

Wekerle, G. (2000a) "From Eyes on the Streets to Safe Cities," *Places* 13(1): 44–9.

—— (2000b) "Women's Rights to the City: Gendered Spaces of a Pluralistic Citizenship," in E. Isin (ed.) *Democracy, Citizenship and the Global City*, London: Routledge, pp. 203–17.

Wekerle, G. R. and Whitzman, C. (1995) *Safe Cities: Guidelines for Planning, Design and Management*, New York: Van Norstrand.

15 Scalar dialectics in green

Urban private property and the contradictions of the neoliberalization of nature

Nik Heynen and Harold A. Perkins

Death and the public reincarnation of Milwaukee's forest: from cathedrals of green to the mundane

Walking through Milwaukee's affluent Eastside, it is difficult to imagine that the presence of its well-maintained trees has been produced by anything but decidedly local processes. However, it is more likely that the distribution of these trees is influenced by, and influences, global socionatural processes shaped through the edicts of neoliberal global capitalism. It is also likely that the urban forests of thousands of other cities across the planet are produced through correspondingly contradictory processes. One need travel only 20 blocks west from the Eastside to realize that Milwaukee's "inner-city" conspicuously lacks the presence of mature and well-cared for trees. It is not their presence, but in stark contrast, their absence that provides the most striking example of the impact of the neoliberalization of urban space on something as seemingly obscure and mundane as the distribution of trees.

In the 1960s and 1970s, Dutch elm disease ravaged Milwaukee's urban forest. The ecological catastrophe led to the loss of more than 200,000 trees. Rigorous government investment has restored 99 percent of the city's public forest. However, only about 4 percent of Milwaukee's urban forest is located on publicly maintained city land. (Heynen *et al.* 2006b). Thus, while Milwaukee has historically had one of the best managed and funded urban forestry departments in the U.S., it contributes little to the total stock of urban trees in the city, creating a "green illusion."

Despite the fact that contemporary Milwaukeeans rank the need for "urban landscape service" provision only behind police and fire service, there have been cutbacks in funding for urban forestry programs since the late 1990s (City of Milwaukee 1996). This is despite an increasingly neoliberal political economy that requires interurban competition for capital investment, which, in turn, necessitates the need to reforest the city in order to increase the quality of life for prospective employers and employees. The Metropolitan Milwaukee Association of Commerce (MMAC), with its claim that people should move to Milwaukee because "it's naturally beautiful," illustrates this socionatural strategy. But without large-scale investment in the urban forest, Milwaukee boosters may soon have to come up with another catchy slogan. The potential ramifications for public ecologies based on Milwaukee's past and its likely future, are ominous. Two examples help to illustrate the city's – and by extension, the planet's – relational dependency on healthy green infrastructures: the urban forest's capacity to soak up surface waters to prevent or reduce flooding and its ability to decrease energy consumption, and thus carbon emissions, by moderating temperature extremes.

Relational importance of Milwaukee's urban forest and public ecologies in "global cities"

In May 2004, the Milwaukee Metropolitan Sewerage District (MMSD) intentionally released 4.6 billion gallons of stormwater runoff mixed with raw sewage directly into Lake Michigan to avoid backing up the "Deep Tunnel," the city's sanitary storage system. This was done to prevent basement damage during intense storms that dumped between 2.72 and 4.93 inches of rain across metropolitan Milwaukee, inundating the Deep Tunnel. The mix of impervious urban surface with deficient urban tree coverage is a fundamental socionatural contradiction that led to the saturation of the Deep Tunnel.

Since pollution released into the lake is not confined to the Milwaukee shoreline, the state of Illinois considered suing Milwaukee over potential exposure to pollution. The Lake Michigan Federation, a powerful citizens' group comprised of people from Wisconsin, Illinois, Indiana and Michigan, has used assorted legal procedures and negative publicity to pressure MMSD to reduce overflows. Such publicity makes "selling" Milwaukee problematic, since it negates the notion that "it's naturally beautiful." The same trees throughout Milwaukee that intercepted rainfall in May 2004 also sequester approximately 1,677 tons of carbon annually (American Forests 1996). Carbon dioxide is the most abundant greenhouse gas and a by-product of the combustion of fossil fuels. Increased numbers of trees appropriately planted near homes in Milwaukee would help lower household energy consumption, and therefore carbon dioxide production, by buffering homes against seasonal temperature extremes.

Forest ecologists, mindful of increasing global deforestation and the potential for thousands of "tree spaces" in cities that could reduce greenhouse gas emissions, have begun to press for a more global perspective on the importance of urban forests (Nowak 1993). Increasing Milwaukee's canopy cover to 40 percent could sequester up to 5,000 tons of carbon annually (American Forests 1996). Substantial increases in urban forests worldwide would also help stem the loss of habitat that supports biodiversity.

The potential for significant reductions in carbon emissions and increases in biodiversity in Milwaukee – just one of many cities in the planetary system of global cities – provides examples of the scalar socionatural relations connected to the commodification of urban forests. The process of commodification, in turn, has local, regional, and global impacts. Although urban land use comprises a small proportion of the planet's surface, maximizing urban forestry in all global cities would provide substantial global ecological benefits.

If urban institutional structures responded better to the dialectical nature of urban metabolic processes, more trees would likely have been planted during Milwaukee's history, and perhaps more could be planted in the city's future. However, historic trends do not bode well for that prospect. Flight from the city by wealthier residents has led to greater income inequality within the metropolitan area and higher rates of housing rentership by low-income residents living in the central city. In addition to the transfer of lands from public to private ownership, the process by which Milwaukee's urban forest is planted and maintained is moving into the private sector. Such a scenario is likely to diminish the size of the urban forest, as a shrinking proportion of the city's population is able to afford to plant trees.

Global/local contradictions and the neoliberalization of "small g's" everywhere

Scalar dialectics is useful in understanding the impacts of neoliberalization on global and local environments. By substituting decidedly spatial – specifically scalar – "processes" in place of fetishized "things" or "places," scalar dialectics link the past and possible futures to spatial "relations," tying them to all the relations that form material realty (see Ollman 2003). Thus, scalar dialectics elucidate the processes and relations that shape urban environments, which, in turn, contribute to the production of regional, national and global environments.

In order to understand the relations that produce the scalar contradictions of neoliberal capitalism within Milwaukee's urban forest, we need to acknowledge the narrow way in which "Global Cities" are considered. There exists a socially produced hierarchy of global cities, each commanding a certain proportion of the global totality of wealth and power. The "big g" cities of New York, London, and Tokyo are at the top of this hierarchy. However, as Luke (2003) points out, more attention must be paid to small and mid-sized urban centers, like Milwaukee, across the earth. Although these cities are not as large in individual population, extent or impact as "big g" global cities, "small g" cities collectively have many times their population and impact. Furthermore, "small g" cities are global, because they are home to a new majority of the world's human population. How they evolve is structured by, and in turn affects, the trajectory of capitalist development and expansion across the planet.

Fewer and fewer urban locations across the planet are immune to the effects of neoliberal capitalism, which has spread unfettered to cities once less vulnerable to its effects. Unique social, physical, and imagined processes have always shaped the evolution of individual urban centers. Yet capital's habit of annihilating space with time is making cities increasingly interconnected, erasing their differences. As urban networks increase in size and extent, the facilitation of capital's never-ending quest for the expansion of value is recognized and made all the more powerful, as well as contradictory. Although urban areas may experience devaluation necessary to the capitalist endeavor, there are always cities willing to provide precious financial subsidies as part of their relentless interurban competition for investment.

The incursion of markets associated with capitalist institutions and corporations is not a natural process, but rather the continuation of Western imperialism under neoliberal capitalism. As marketization and the privatization of property spread, aided by the mantra of inevitability, urban ecological crises increase in destructive and spatial/scalar extent. As Brenner and Theodore (2002: 23) suggest, neoliberal localization results in the creative destruction of urban public services and infrastructures. Public services and publicly owned infrastructures are in essence "destroyed" as they are turned over to the private sector for management. The moment of creativity involves the production of "privatized, customized, and networked urban infrastructures intended to (re)position cities within supranational capital flows." The emergence of new markets for service to public infrastructure in no way guarantees investment in urban trees, thus jeopardizing green infrastructures formerly underwritten by the state.

Quantity/quality relations provide a context for understanding the contradictory scalar relations associated with the role that urban areas like Milwaukee play in global public ecologies. Engels (1969) discussed quantity/quality relations to explore contradictory relations inherent to capitalism. Ollman (2003) suggests: "What is called quantity/quality is a relation between two temporally differentiated moments within the same process." We recognize the temporal importance of this relation for elucidating the contradictory nature of capitalism but suggest this relation can also be mobilized to more systematically understand scalar contractions of urban forestry in Milwaukee. Ollman goes on to say:

> Every process contains moments of before and after, encompassing both build up (and builddown) and what that leads to. Initially, movement within any process takes the form of quantitative change. One or more of its aspects – each process being also a relation composed of aspects – increases or decreases in size or number. Then, at a certain point – which is different for each process studied – a qualitative transformation takes place, indicated by a change in its appearance and/or function.

In order to comprehend these processes, qualitative changes often necessitate new ideas, conceptualizations and knowledge formation. This new knowledge is necessary in order to recognize what the process has become. Quantity/quality relations can help to better articulate the contradictory processes inherent in the causes and consequences of global/local deforestation. For instance, when the quantity of something changes, say the forest shrinks from 100 trees to 50 trees, a quantitative change has occurred. In this case, we can still discuss the importance of the forest, because the quantitative change has not altered the fundamental nature of what we are discussing. However, if the same forest diminishes by 99 trees, we can no longer talk about a forest, because the basic qualitative nature of what we are discussing has changed. What remains is but a tree.

Within the context of global/local forestry, the increased relative importance of urban forest ecology within academia, planning and politics relates to wide-ranging quantitative relations resulting from increased global deforestation (Nowak and Dwyer 2001). As more and more trees are cut down in the Amazon Basin, Central Africa, and Southeast Asia, their relative scalar importance increases (Heynen 2003). The socionatural relations in question remain the same; however, the relative scarcity of global trees exacerbates environmental crisis and should lead to increased and diversified knowledge formation. Consequently, interrelations can be drawn between the diminishing urban forest due to factors such as attrition or Dutch elm disease and increased urban environmental problems related to stormwater runoff or heat island effect.

Current trends regarding global/local forest quantity/quality relations conform to the definition of a scalar ecological contradiction. The contradiction is inherently based on our inability/unwillingness to generate the knowledge formation necessary to respond to global/local deforestation. As environmental resources produced in discrete spatial locations, urban trees provide opportunities to gain insight into these both specific and more general ecological contradictions within private property relations.

Scalar contradictions in relation to consumption fund underinvestment

The urban built environment serves as fixed capital for the renewal of production. It can also be consumed to sustain the existence of labor, therefore contributing again to the production process. Harvey (1989: 36) states "the built environment can be divided conceptually into fixed capital items to be used in production (factories, highways, railroads, and the like)." The consumption fund is also the circuit of capital that contains most urban ecological resources, such as urban forests, that are vital to the health of urban systems in general. Historically, many artifacts within the consumption fund have been underwritten and maintained publicly. Current processes of neoliberalization, steeped in free-marketeering rhetoric, have increased the apparent legitimacy and inevitability of the right to own private property. This perspective is diametrically opposed to all other forms of ownership including public, state or collectively owned forms of property. Not surprisingly, state (read municipal) investment is being directed increasingly toward private development projects at the expense of publicly owned infrastructure.

However, investment in publicly owned portions of the consumption fund is critical to the continuation of production conditions. As Harvey (1989: 80) states, " ... most difficult of all to grasp is the manner in which production and consumption relate so that each of them creates the other in completing itself, and creates itself as the other." As Harvey (1999: 65) further suggests, expenditure on consumption fund items depends upon collective investment by the state: "A general condition for the flow of capital into the secondary circuit is, therefore, the existence of a ... state willing to finance and guarantee long-term, large-scale projects with respect to the creation of the built environment." In other words, by taxing capital and spending revenues on publicly owned property, the state compensates for individual capitalists' reluctance to invest in infrastructures that renew the cycle of production and accumulation within urban centers. Additionally, the ability of many people to benefit from consumption fund amenities is dependent upon continuous forms of state investment. Retrenchment of investment by the public sector in collectively owned green infrastructures serves to discourage low-income residents from consuming public ecologies as they are privatized under neoliberal capitalism.

The presence of trees on Milwaukee's Eastside in contrast to a lack of trees within its inner city demonstrates the divergent ability of Milwaukee's residents to produce and use the urban forest. Accordingly, the "ownership society," as extolled by neoliberals, results in private investment in urban forest located on private property. Simultaneously, decreased responsibility by the state for the production and stewardship of forests on urban public areas diminishes the ability of poorer citizens to live with urban forest amenities, which were formerly underwritten by the state.

Because urban forests are socially produced, urban political economy increasingly dictates that they are produced in accordance with market forces that operate at various scales. While the decisions to disinvest in consumption fund items like the urban forest are often made locally, they are impacted directly by forces operating at regional and global scales. It is therefore possible to connect local fiscal crises to economic restructuring under neoliberalism that direct localities to invest in private projects tied to global investment capital, rather than a public good such as Milwaukee's urban forest. Paradoxically however, the city's ability to attract more global investment capital decreases as production of commodities for the renewal of labor and green infrastructure diminishes. This contradiction of urban neoliberalization negatively affects the integrity of the city's urban environment, not only for its low-income residents, but also for potential investors. By scalar extension, global public ecologies are further decimated as similar processes affect cities across the planet.

Local contradictions of the neoliberalization of Milwaukee's urban forest

Milwaukee's development is embedded within the rise of industrial capitalism in the U.S. during the nineteenth and twentieth centuries. Largely a working-class city, Milwaukee's laborers have a long history of organized opposition to exploitative city politics. Working and living conditions within Milwaukee toward the turn of the nineteenth century resulted in upheaval among the working classes. Tired of abuse by capitalists and unsanitary living conditions, Milwaukeeans elected their first Socialist mayor, Emil Seidel, in 1910. The "Sewer Socialists" ran on a platform aimed at eliminating cronyism within city government and repairing and creating badly needed public infrastructure throughout the city. When the Socialists took charge in 1910, they immediately reorganized the Department of Public Works to accommodate the public needs of the city. By halting cronyism, and thereby reducing total annual expenditures, the first Socialist administration was able to rebuild the city's decrepit infrastructure, create employment, and improve the social and material quality of Milwaukee.

The Great Depression triggered waves of collective investment in public infrastructure across the U.S. Nationally, Roosevelt's New Deal legislation signaled increased investment by the state into forms of property publicly or collectively owned. But increased federal expenditure on urban green infrastructure was preceded by an already dedicated interest to promote collectively owned property in the form of local parks and trees by Milwaukee Socialists. Under Socialist mayor Daniel Hoan, thousands of unemployed men were working to expand Milwaukee's parks and plant more trees, leading to the creation of one of the "greenest" U.S. cities by the end of the Depression. Hoan's (1936) motivation for such rigorous public investment was that "workers are interested in parks and playgrounds, the health-building and character-building centers of cities." Thus the Socialist party made Milwaukee a city world-renowned for the quality and extent of its built/green public infrastructures during the first half of the twentieth century.

Collectively, these projects not only spurred spending by increasing employment but also contributed to the sense of place and standard of living for Milwaukee residents. Milwaukeeans have historically loved their trees; it is a part of their shared identity (American Forests 1996). As a result of Milwaukeeans' appreciation of their green space, the city has one of the largest and oldest urban forestry programs in the U.S. today. Nationally recognized for doing so, it has a long and important history in fostering and maintaining the trees that city residents demand.

While the results of these capital investments in infrastructure are publicly owned, they do not stand in contradiction to capitalist accumulation. The construction of amenities such as parks and boulevards, along with the planting of trees, represented a significant investment for improving Milwaukee's ability to compete for global capital investment. These expenditures are precisely what has given Milwaukee a reputation as a good place to live and do business, facilitating renewed accumulation and leading to its popular portrayal as the heart of middle-class America in such television shows as *Happy Days* and *Laverne & Shirley.*

Fiscal austerity, in part a result of an increasing fear of raising taxes, is the current mantra accepted by politicians at all levels of government and on both ends of the mainstream U.S. political spectrum. Milwaukee has lost millions of dollars of revenue-shared funds from the state of Wisconsin in the mid-2000s, which has directly affected its parks and forestry departments. In conjunction with reduced taxation, neoliberal policy encourages increased privatization of property for the sake of the continued dismantling of the state's influence over property markets. Consequently, to save money, the parks and forestry departments face budget reductions while more publicly owned property in Milwaukee has become privatized. These developments present a conundrum for the city because in order to attract investment capital, Milwaukee must be competitive. But in order to compete effectively, major expenditures on publicly owned infrastructure are necessary.

Therefore, it seems that under a neoliberal mode of governance dictating fiscal austerity, the relationship between production and consumption within the capitalist political economy is at risk (Heynen and Robbins 2005). Thus, we have continued consumption of built and natural urban infrastructure in the form of parks and trees, while their production and maintenance are diminished. Diminished expenditure on production conditions in urban environments represents O'Connor's (1998) second contradiction of capital, whereby underproduction of production conditions result in fiscal and ecological crisis. This contradiction within capitalism not only diminishes the vitality of the capitalist system itself, but threatens the continued privatization of public resources. As O'Connor suggests, new social forms of crisis management under capitalist relations may be necessary in the future if cities like Milwaukee are to recover from neoliberal underproduction of the conditions necessary to renew accumulation.

Regulations adopted by the City of Milwaukee for urban trees planted on private property only authorize the removal of trees that are considered a nuisance. Diseased trees pose risks to healthy trees; the city gives itself the right to cut down diseased trees on private property if the tree in question threatens the health of surrounding trees. While this policy may seem proactive, a deeper examination of what the regulations fail to address is more telling. The planting of trees on private property has always fallen under the notion of self-determination by the property owner. Consequently, city regulations state nothing about the replanting of a tree lost to disease or damage, so the city is powerless to mandate replanting. This omission means that the City of Milwaukee is only allowed to react to threats to its forest on an individual basis, and only after the disease or damage has occurred.

Liberal underpinnings in urban forestry management in Milwaukee were most readily apparent during the Dutch elm disease crisis that swept through the city between 1965 and 1980. City residents expressed much regret over the loss of stately boulevard elms that lined city streets, forming cathedral-like archways. However, tens of thousands of trees were lost on private property as well. Under city regulations, the forestry department replanted trees lost to Dutch elm disease on city property, but not on private property which is the responsibility of the individual property owner. According to the Milwaukee forestry department, 6,153 trees were cut down on private property between 1994 and 2002. While there are no data on how many of those trees were replanted, forestry department staff says the number is far fewer.

This is important because self-interested property owners may not consider the planting of a new tree on their property to be to their benefit, let alone the benefit of the city/world as a whole. If the city does not replace lost trees on private property and the ratio of public to private property decreases, Milwaukee's ability to provide a better quality of life and compete to attract investment capital diminishes. This affects local socionatural relations and accumulation regimes as well as having regional and global impacts, because the loss of urban trees cumulatively affects climate, air quality, and the quality of life, as well as many other ecological processes.

Privatization and the uneven reforestation of Milwaukee

Milwaukee's urban forest continues to suffer because of the push within the state to diminish expenditures across a range of social service and infrastructure provisions via privatization. Efforts today to replant trees on private property in Milwaukee fall to *Greening Milwaukee*, a not-for-profit, quasi-private institution that depends largely upon donations from private sources and grants from public agencies in order to fulfill its mission. The intrinsic value of trees is meaningless according to the terms of neoliberalization, so Greening Milwaukee justifies its tree planting goals through market valuations of the urban forest.

Trees found on private residential and private commercial/industrial property account for approximately 55 percent of the total urban forest resource in Milwaukee. American Forests (1996) suggests that trees in Milwaukee reduce total stormwater runoff volume by 5.5 percent and reduce peak flow by 9.4 percent. The direct summer energy savings from Milwaukee's urban forest was valued at $650,000 during 1996 alone. American Forests also estimates that adding one mature tree in the right location at each home (on the West or East side and shading the air conditioner) would boost that savings to more than $1.5 million a year. Succumbing to market logic as their only means of defense, the champions of Milwaukee's urban forest pitch these numbers to potential donors, policy makers, and the public at large.

Despite the need for enhancing urban forestry on the currently expanding proportion of private property in Milwaukee, the future of the institution entrusted with that task is not guaranteed. Although it receives grant monies from public entities like the City of Milwaukee and Wisconsin Department of Natural Resources, Greening Milwaukee does not have a secure source of annual public funding. Because of budget cutbacks, the organization must depend more on private donations.

Beyond its precarious funding structure, there are also questions about whether a quasi-private, not-for-profit organization can address the loss of urban trees on private property in the city, given that a large proportion of the city's low-income renters may not have access to adequate transportation, the money, or permission to plant trees on rental property. Approximately 55 percent of all Milwaukeeans rent their homes. Unfortunately, housing status heavily influences participation in Greening Milwaukee, particularly its "Adopt-A-Tree" program (Perkins *et al.* 2004). Thus, in 2002, renters comprised only 11 percent of all participants in the program, versus the remaining 89 percent who were homeowners. While, Greening Milwaukee will deliver trees to those who cannot pick them up, only 15 trees were planted on rental properties by "Adopt-A-Tree" throughout the entire city in 2002.

In an increasingly neoliberalized environment, the only people able to address the city's ecological needs are those wealthy enough to own their homes. We suggest that greater neoliberalization will likely lead to an increasingly uneven urban ecology in Milwaukee, relegating lower income groups to areas that lack trees. Thus, while Greening Milwaukee is a local organization operating within a maze of local private property relations and neoliberal governance structures, the terms of its existence also inevitably contribute in some measure to the degradation of the global ecosphere.

Conclusion

To conclude by suggesting that neoliberal capitalism is rife with contradictions would border tautology. However, this chapter suggests there is still more work to be done in understanding how neoliberalization affects socionatural relations at all scales. Trees, one of the most visible natural components of the urban environment, help Milwaukeeans identify with their city and help boosters advertise the city as a good place to live and work. Investment capital placed into the consumption fund, in the form of urban trees, must be expended if the city hopes to attract capital from investors outside and within Milwaukee. Therefore, the production of urban ecology must be enhanced in order for Milwaukee to compete effectively with other cities vying for investment capital.

Greening Milwaukee illustrates that neoliberal forms of urban reforestation require intense scrutiny. Such organizations have the potential to exclude the majority of its citizenry based on socioeconomic factors like housing ownership. While this process conforms to the inequality produced through capitalism, illustrating these contradictions within the context of the social production of urban environments helps to bolster the argument against the continued neoliberalization of nature. And this, of course, has potentially enormous scalar consequences, ranging from impacts on individuals living in urban areas to global ecological processes.

As the neoliberalization of urban areas across the planet proliferates, contradictions inherent in neoliberal capitalism will continue to degrade urban environments (Heynen *et al.* 2006a). The planetary network of cities, or "small g" global cities, continue to be the destination for large numbers of immigrants and are expanding rapidly in area and socionatural importance. Urban environmental crises, and by extension planetary environmental crises, will continue to grow in breadth and depth. More emphasis and political pressure need to be placed on the role that the city or state takes in planting and maintaining trees on privately owned property.

References

American Forests (1996) *Urban Ecological Analysis for Milwaukee, Wisconsin*, Washington, DC: American Forests.

Blomely, N. (2004) *Unsettling the City: Urban Land and the Politics of Property*, London and New York: Routledge.

Brenner, N. and Theodore, N. (2002a) *Spaces of Neoliberalism: Urban Restructuring in North America and Western Europe*, Oxford: Blackwell Publishing.

—— (2002b) "Cities and the Geographies of 'Actually Existing Neoliberalism'" in N. Brenner and N. Theodore (eds.) *Spaces of Neoliberalism: Urban Restructuring in North America and Western Europe*, Oxford: Blackwell Publishers.

City of Milwaukee Department of Public Works Forestry Division (1996) *Comprehensive Boulevard Study*, Milwaukee, WI: City of Milwaukee.

Engels, F. (1969) *Anti-Dühring: Herr Eugen Dühring's Revolution in Science*, Moscow: Progress Publishers.

Gurda, J. (1999) *The Making of Milwaukee*, Brookfield, WI: Milwaukee County Historical Society.

Harvey, D. (1989) *The Urban Experience*, Baltimore, MD: The Johns Hopkins University Press.

—— (1999). *Limits to Capital*, London: Verso.

Heynen, N. (2003) "The Scalar Production of Injustice within the Urban Forest," *Antipode: A Journal of Radical Geography* 35: 980–98.

Heynen, N., Kaika, M. and Sywngedouw, E. (2006a) *In the Nature of Cities: Urban Political Ecology and the Politics of Urban Metabolism*, London: Routledge.

Heynen, N., Perkins, H. A. and Roy, P. (2006b) "The Political Ecology of Uneven Urban Green Space: The Role of Private Property, Class and Race/Ethnicity in Producing Environmental Inequality in Milwaukee," *Urban Affairs Review* 42(1): 3–25.

Heynen, N. and Robbins, P. (2005) "The Neoliberalization of Nature: Governance, Privatization, Enclosure and Valuation," *Capitalism, Nature, Socialism* 16(1): 5–8.

Hoan, D. W. (1936) *City Government: The Record of the Milwaukee Experience*, New York: Harcourt, Brace and Company.

Luke, T. W. (2003) "Global Cities vs. 'global cities:' Rethinking Contemporary Urbanism and Public Ecology," *Studies in Political Economy* Spring: 11–33.

Marx, K. (1964) *Economic and Philosophical Manuscripts of 1844*, trans. Martin Milligan, New York: International Publishers.

—— (1990) *Capital*, vol. 1, trans Ben Fowkes, London: Penguin Books in association with New Left Review.

McCarthy, J. and Prudham, S. (2004) "Neoliberal Nature and the Nature of Neoliberalism," *Geoforum* 35: 275–83

Marx, K. with Engels, F. (1998) *The German Ideology*, New York: Prometheus Books.

Nowak, D. J. (1993) "Atmospheric Carbon Reduction by Urban Trees," *Journal of Environmental Management*, 37: 207–17.

Nowak, D. J. and Dwyer, J. (2001) "Understanding the Benefits and Costs of Urban Forest Ecosystems," in J. Kuser (ed.) *Urban and Community Forestry in the Northeast*, New York: Kluwer.

O'Connor, J. (1998) *Natural Causes: Essays in Ecological Marxism*, New York: Guilford Press.

Ollman, B. (2003) *Dance of the Dialectic: Steps in Marx's Method*, Urbana, IL: University of Illinois Press.

Perkins, H. A., Heynen, N. and Wilson, J. (2004) "Inequitable Access to Urban Reforestation: The Impact of Urban Political Economy on Housing Tenure and Urban Forests," *Cities*, 21: 291–9.

Sampson, R. N. (1989) "Needed: A New Vision for Our Communities," in G. Moll and S. Ebenreck (eds.) *Shading Our Cities: A Resource Guide for Urban and Community Forests*, Washington, DC: Island Press.

Swyngedouw, E. (1996) "The City as a Hybrid: On Nature, Society and Cyborg Urbanization," *Capitalism, Nature, Socialism*, 7: 65–80.

16 Neoliberalism and environmental justice policy

Ryan Holifield

Introduction

It took a special kind of neoliberalism to make room for the concerns of the US grassroots environmental justice movement. Concerns that non-white and poor communities bore a disproportionate share of the nation's environmental hazards found a receptive ear in the Clinton administration, which in the 1990s replaced the aggressive neoliberalization of the Reagan era with what might be called a "kindlier, gentler" neoliberal project. Although Clinton's famous Executive Order No. 12898 ("Federal Actions to Address Environmental Justice in Minority Populations and Low-Income Populations") is by no means a statement of neoliberal doctrine (Clinton 1994), it nonetheless lays the groundwork for an approach to addressing environmental justice which was compatible with what Peck and Tickell (2002) have called "roll-out" neoliberalism.

This chapter addresses three primary questions. First, how did the approach to environmental justice sketched out in the policies and programs of the Clinton administration potentially advance the project of consolidating and "deepening" neoliberal hegemony? Second, how did the administration's environmental justice policy affect decision-making practices among "street-level" implementers? Third, did it affect these decision-making practices in ways that could potentially transform the conditions of primary concern for environmental justice activists – uneven distributions of health risk and a lack of political and economic power? The chapter explores these questions through a case study of the US Environmental Protection Agency's (EPA) Superfund hazardous waste site remediation program, which linked the Clinton administration's approach to environmental justice with its initiatives for revitalizing inner cities and other economically struggling areas.

"Deep neoliberalization" and environmental justice

I argue that the Clinton administration's approach to environmental justice constituted part of what Jessop (2002) identifies as a *neocommunitarian* strategy for extending and sustaining the project of neoliberalization. Instead of attempting to *redistribute* environmental risk more equitably – a conceivable collectivist approach to environmental justice – this strategy revolved around efforts to *empower* and *build trust* in "environmental justice communities." At the same time, it involved technocratic attempts to *standardize* these communities using geographic information systems (GIS) analyses.

At the ideological core of neoliberalism is a utopian commitment to extending the competitive relations of the market as far as possible, keeping state intervention to a minimum. "Actually existing neoliberalism" consists of restructuring projects designed to realize this utopian vision (Brenner and Theodore 2002). In the US and the UK during the 1980s, these projects – as part of what Peck and Tickell (2002) call "roll-back" or "shallow" neoliberalization – emphasized gutting regulations and dismantling institutions of the Keynesian welfare state. In contrast, the projects of "roll-out" or "deep" neoliberalization of the 1990s focused on constructing new institutions of federal intervention to consolidate and deepen neoliberal hegemony (Jessop 2002).

One strategy associated with "deep neoliberalization" has been the selective appeal to "community" and non-market values as "flanking mechanisms" for neoliberal projects (Brenner and Theodore 2002, Jessop 2002, Peck and Tickell 2002). Jessop (2002: 463) identifies this strategy as *neocommunitarianism*, which emphasizes

> the link between economic and community development, notably in empowering citizens and community groups; the contribution that greater self-sufficiency can make to reinserting marginalized local economies into the wider economy; and the role of decentralized partnerships that embrace not only the state and business interests but also diverse community organizations and other local stakeholders. The neocommunitarian strategy focuses on less competitive economic spaces (such as inner cities, deindustrializing cities, or cities at the bottom of urban hierarchies) with the greatest risk of losing from the zero-sum competition for external resources ... It aims to redress the imbalance between private affluence and public poverty, to create local demand, to re-skill the long-term unemployed and reintegrate them into an expanded labor market, to address some of the problems of urban regeneration (e.g. in social housing, insulation, and energy-saving), to provide a different kind of spatiotemporal fix for small and medium-sized enterprises, to regenerate trust within the community, and to promote empowerment.

The deployment of a neocommunitarian strategy made the idea of environmental justice, often invoked in grassroots activist discourse as a challenge to market-driven environmental policy, compatible with processes of neoliberalization. Although neocommunitarianism obviously did not embrace the more radical demands of the environmental justice movement ("Principles of Environmental Justice" 1991), it directly addressed numerous themes animating environmental justice activism, such as community empowerment, citizen involvement, and economic self-sufficiency. First, although it ignored demands for full democratic participation in decision-making, the Clinton administration made "public participation" central to environmental justice policy. The executive order conceived this participation primarily in terms of improved public relations – making the agencies' decisions "accessible" and allowing the public to submit recommendations – and carefully managed community involvement.

Second, the Clinton administration treated environmental injustices as opportunities to create private-sector jobs and to stimulate new investment in neglected communities. Under Clinton, EPA established grant programs for EJ communities, expanded or created training programs for jobs cleaning up hazardous waste, and offered incentives to redevelop brownfields – hazardous waste sites that failed to qualify for remediation under Superfund, often located near economically disadvantaged neighborhoods.

Along with public participation and economic development, the third major component of the Clinton administration's approach to environmental justice was data analysis (Foreman 1998). Executive Order No. 12898 called for federal agencies to "collect, maintain, and analyze information" on demographics and levels of risk, which would then serve as the potential basis for verifying claims of environmental injustice and defining the boundaries of the standardized "EJ community." By making environmental injustice subject to calculation, measurement, and mapping, the activity of standardizing the "environmental justice community" could thus potentially help consolidate a technocratic basis for federal environmental justice policy decisions. "EJ communities" would qualify for targeted grants and programs, but only if state-controlled technical analysis verified their "EJ" status.

Environmental justice and the Superfund program

Superfund is the more familiar name for the Comprehensive Environmental Response, Compensation, and Liability Act (CERCLA) of 1980, renewed by the Superfund Amendments and Reauthorization Act (SARA) of 1986. The original law gave EPA authority and funding to identify and clean up the country's worst abandoned hazardous waste sites. It covered two kinds of cleanup: (1) time-critical responses to emergencies, and (2) more expensive long-term remedial actions, designed to clean up sites more thoroughly and permanently. EPA implements the latter only at sites on the National Priorities List (NPL), a subset of all hazardous waste sites identified in EPA's Comprehensive Environmental Response, Compensation, and Liability Information System (CERCLIS) database.

In remedial actions, decisions about how to clean up individual sites usually rest with remedial project managers (RPMs), street-level bureaucrats trained as engineers or scientists. RPMs are constrained by federal and state guidelines, which set a range of risk levels deemed to be acceptable.[1] Still, they always have multiple options for how to clean up any individual site, and EPA requires them to seek the least expensive solution that achieves acceptable levels of risk. The range of options might include sealing the waste under a clay cap, incinerating it, neutralizing it on site, or hauling it away to be treated elsewhere. Each of these possibilities costs a different amount of money, remains effective for a different length of time, and reduces health risks to a different degree.

Two aspects of the remedial decision-making process have emerged as environmental justice concerns: the pace of clean-up and the stringency and permanence of selected remedies. Empirical studies have reached conflicting conclusions about whether Superfund sites are remediated more slowly or less permanently in predominantly low-income and non-white communities (Lavelle and Coyle 1992; Zimmerman 1993; Hird 1994; Gupta *et al.* 1995; Anderton *et al.* 1997; Viscusi and Hamilton 1999). But since all of these studies analyze data from the era before Executive Order No. 12898, they cannot address the issue of how the Clinton administration's environmental justice policy might have affected these conditions.

Environmental justice policy and remedial practice in EPA Region 4

How did efforts to calculate the disproportionate effects of environmental risks and define the boundaries of "EJ communities" affect the way remedial personnel made decisions? How did EPA's environmental justice policies affect the speed of the process and the redistribution of risk? Did they lead remedial personnel to consider environmental justice as an explicit criterion in managing risk, or did they alter the management of remediation and distribution of risk in less direct ways? Finally, how did the Clinton administration's neocommunitarian emphases on public participation and economic opportunity influence the decisions of remedial personnel? The following section addresses these questions through a case study of the Superfund program in EPA Region 4, which includes eight southeastern states (Alabama, Florida, Georgia, Kentucky, Mississippi, North Carolina, South Carolina, and Tennessee) and is often considered to be the source of the grassroots environmental justice movement (Bullard 1994).

Remediation and the normalization of EJ communities

The heart of EPA Region 4's *Interim Policy to Identify and Address Potential Environmental Justice Areas* (1999) is a set of procedures for a GIS-based demographics analysis.[2] The protocol for the analysis instructs managers to determine, by comparing census data on race and income in the target area to state-specific minority and low-income "thresholds," whether or not each "target area" (defined by a circular buffer zone drawn around a regulated site) is a "potential EJ area." In theory, the analysis gives managers a scientific tool to verify a community's self-identification as a victim of environmental injustice, an "actual" EJ community – and thus to determine "objectively" whether the community legitimately qualifies for targeted neocommunitarian grants and programs. The *Interim Policy* emphasizes that the analysis is "critical" to managing environmental justice concerns (1999: 6), characterizes it as "the *necessary* tools to *properly* define a potential EJ area of concern" (1999: 3; emphasis added), and insists that prior to remedial activities, "GIS demographics and low-income data *must be obtained* to make an *accurate* determination" of EJ status (1999: 12; emphasis added). An attorney I interviewed contended that "*In the near future, with new census data, self-identification may not be allowed.*"[3]

However, the GIS analysis suffers from two limitations. First, it could not offer managers a methodology for determining "actual" EJ communities, because the EPA had no standard for calculating "disproportionate effects." Since studies have demonstrated that whether particular populations suffer disproportionate effects of environmental risk varies depending on the scale and resolution of analysis (Cutter *et al.* 1996; Kurtz 2002), it would appear impossible (or at least highly improbable) that EPA could reduce the question of which communities are "true" EJ communities to a technical one. The selection of a standard scale and resolution of analysis would ultimately remain a policy decision and a question of judgment – thus inevitably open to legal and political contestation.

Second, even the protocol for determining "potential EJ communities" or "areas" leaves a great deal to the manager's judgment. Since neither "community" nor "area" corresponds to a precise scale, "the remedial project manager (RPM) should determine the area of the demographic analysis based upon his/her knowledge of the site and sampling results" (1999: A32). Not only are site boundaries often ambiguous and contested, but the indicators for "low-income" available to the manager – below $15,000 income or below the poverty line – are also ultimately arbitrary. As a consequence of these limitations, the *Interim Policy* must concede that managers should treat as EJ cases those communities who "self-identify" as such.

Despite the *Interim Policy*'s characterization of the EJ analysis as "critical," the RPMs I interviewed unanimously maintained that they do not use it to verify claims of environmental injustice. Some reported routinely analyzing demographics as part of the Community Relations Plan required for all Superfund remediations, but they explained that they did not connect these analyses to the designation of communities as "EJ cases." Others suggested that they do not conduct the analysis at all.

> From the perspective of the project manager, the demographics analysis doesn't make any difference. We're going to conduct our business in the same way regardless.
>
> (Interview, RPM #1)

I'm not involved with the identification of sites as EJ sites. I work on the sites that are assigned to me ... Some other entity identifies the sites as EJ sites. There might be an asterisk by the site, or perhaps not – maybe our not identifying them is a problem. I don't know who made these decisions.

(Interview, RPM #3)

In general, participants' comments suggested that EJ cases were simply those in which the local community invoked environmental justice as a political issue. As one RPM put it: "*If a community declares themselves an EJ community, we treat them as one*" (Interview, RPM #5). Another RPM told me that at his most controversial EJ site, "*it would have been inappropriate to prove or disprove*" the community's self-identification using the GIS analysis:

The nature of the complaints makes [his primary EJ case] an EJ site. There never has been any formal determination of whether it is an EJ community. It is self-identified ... No one's ever seen it necessary to prove or disprove the identification.

(Interview, RPM #2)

To support his contention, he cited another community a few dozen miles away from his most active site. Although the community's demographics would have qualified it as a potential EJ area according to the GIS analysis, it never raised environmental justice concerns and thus never received the EJ designation.

A community involvement coordinator suggested that the EJ analysis was useful primarily as a means to make communities aware that they are eligible for EJ resources and grants. Instead of a tool for verifying claims of environmental injustice, she described the analysis as a means for making communities – such as low-income white communities – aware that their demographics would qualify them for EJ resources:

Yes, several have self-identified as EJ once they know they qualify ... I have an EJ analysis done; that's the first thing I do. Then I can let them know when I get there. It really depends on what kind of people are there. Some people don't care if they're EJ or not.

(Interview, CIC)

She suggested that demographics make a difference in the remedial process only if communities actively use their demographic status to mobilize for collective political action under the banner of environmental justice.

Environmental justice as the redistribution of risk?

In theory, the EJ analysis could also serve as the basis for managing the redistribution of hazardous waste risk more "justly" or "equitably." But the case study suggests that EPA Region 4 did not deliberately choose more stringent remedies in EJ communities or pursue a "fairer" distribution of Superfund risk along lines of race or class. The *Interim Policy*'s only directive regarding risk in EJ communities is that "the environmental and human health burden experienced by that community should receive heightened scrutiny" (1999: A32). However, the document never elaborates on what "heightened scrutiny" might mean in practice.

All of the remedial project managers who participated in the study insisted that they never based their remedial decisions on the racial, ethnic, or economic composition of the community. In some cases, they sought to refute the accusation that they applied *less* stringent remedies in EJ communities, insisting on their impartiality and their avoidance of preferential or differential treatment:

> In the case of Superfund, some people apply EJ issues by making two allegations: first, that EPA makes decisions on Superfund sites that may be racially motivated – for example, that it may design a less rigorous remedy because the facility is located in minority neighborhood. And second, that EPA drags their feet in minority neighborhoods. I think they're basically crap. I can speak for Region 4, myself, and all of my colleagues: that never is the case. We select a remedy based on science, regardless of who lives around the site. Race does not play a part in my decision. To say we drag our feet, a bunch of crap too. In fact, we have supervisors hounding us: "Finish this thing, get it done!"
>
> (Interview, RPM #1)

But while the *demographics* of race and class may not have played a direct role in remedial decisions, RPMs admitted that the *politics* of race (and to a lesser degree class) sometimes do. One RPM pointed out, for example, that communities who blame their proximity to a Superfund site on racial or economic discrimination might also allege that the EPA's proposed remedy discriminates against them as well:

> They might say we might do clean-ups differently in different places. They make comparisons about remedies. They say, "You cleaned up this community more thoroughly than ours, that's not fair!" Our decisions are never based on the ethnic makeup or economic status of a community, but that concern raised by them will force the agency to respond and defend its decision. The community might ask, "Could you have done anything else?" The answer will always be yes. We'll change the remedy if another one meets the criteria, to fight off allegations that we're discriminating against them.
>
> (Interview, RPM #3)

In short, the case study suggests that while environmental justice policy had no *direct* effect on the distribution of health risk associated with hazardous waste sites, it indirectly affected this distribution by forcing EPA to respond to localized claims of environmental injustice.

Remediation and the empowerment of EJ communities

The Superfund Amendments and Reauthorization Act (SARA) of 1986 contained provisions to ensure "meaningful public participation" in any remedial action, regardless of whether or not the affected community is an EJ case. Most of these appear in the *Interim Policy* under the heading of either "community relations" or "community involvement." While community *relations* activities in the remedial process focus on communication *from* EPA to the people near the site, community *involvement* activities emphasize encouraging affected residents and other stakeholders to communicate their concerns *to* EPA. What makes community relations and community involvement in EJ communities different from other sites? The document suggests that if an individual or group claims to have "EJ concerns," Region 4 personnel should "enhance" community involvement and relations efforts (1999: A32). It also calls for Region 4 to establish local and regional forums "for community input or resolutions," such as Community Advisory Groups (CAGs) and regional EJ Summits (1999: A39–40). However, the directive to "enhance" remains unspecified, and although CAGs were developed with EJ communities specifically in mind, the document points out that the agency could eventually apply them in any affected community.

Still, at least two aspects of the *Interim Policy*'s recommended approach to community relations and involvement in "EJ communities" stand out as distinctive. First, the document encourages RPMs to use additional, specialized staff – Environmental Justice Coordinators – to assist in community involvement. Second, it directs them to make qualified communities aware of EPA grants set aside for EJ communities, such as the EJ Small Grants Program and the State and Tribal Environmental Justice Grants Program. Unlike Technical Assistance Grants (TAGs), which are available at all Superfund sites to support the hiring of a technical advisor, environmental justice grants are available only to groups addressing concerns in minority or low-income communities. In 2001 (the year this research was completed) EPA allocated $1.5 million to the EJ Small Grants Program, one third of which it set aside specifically for Superfund projects. The *Interim Policy*'s Community Involvement section links these EJ grants to the development of political strength, calling them a potentially "effective tool for empowering and expanding the capacity of the community to organize, monitor and play a continuing role in environmental decision-making" (1999: A42). In effect, the EJ grants serve as instruments for the neocommunitarian objective of supporting the political and economic self-sufficiency of marginalized communities.

Did Region 4 remedial personnel "enhance" community involvement and relations efforts in EJ communities? The Community Involvement Coordinator I interviewed claimed that

I treat them all the same. It's my understanding that EJ communities should get more attention, more focus. But we treat them all the same.

(Interview, CIC)

However, the accounts of other participants suggested that communities who identified themselves as EJ cases usually received additional resources and attention, which a Superfund attorney described as "outreach plus":

We are asked to be "extra-available." We do more presentations than we normally do. The community tends to be more active.

(Interview, attorney)

One objective of this enhanced outreach in EJ communities was to restore or build trust in remedial decisions in the face of allegations of discrimination. A white RPM told me that at his most controversial EJ site:

They [the community] will tell you flat out that the way that area got how it is is because of race ... we've taken our cue from the people.

(Interview, RPM #2)

He noted that the high level of racial tension and distrust had caused him "real difficulties with the community" at an earlier stage of the process, but claimed that his relations with the community had improved greatly ever since Region 4 assigned one of its EJ coordinators to the site. He indicated that he now felt more confident that the community would trust him and accept his remedial decision. At other sites, however, deploying an EJ coordinator was not enough. As one supervisor noted,

EJ communities often want an African American RPM, even though [one of the African American RPMs she supervised] is one of the toughest I've got ... It's hard for a white RPM to build that trust, even if a white RPM might provide what they need.

(Interview, RPM #4)

Indeed, a white RPM under her supervision had been replaced by an African American RPM at a controversial site after a predominantly African American community rejected his remedial decision.

In short, the Clinton administration's neocommunitarian policies led EPA Region 4 to make community acceptance of remedial decisions a major priority – something that had not always been the case. Although several participants lamented the "interference" of politics in what they maintained should be a completely objective, scientific process, some indicated that they approved of the Clinton administration's focus on involving communities in order to build trust in the remedial process.

The *Interim Policy*'s guidelines for implementing environmental justice in remedial activities also contain a section that advises remedial personnel on "Responding to Grants/Economic Development-Related Concerns in the Remediation Process." In addition to the various EJ grants, the section cites Worker Training, Community-Based Environmental Protection, and Brownfields programs as "tools for achieving economic development in EJ communities" (1999: A35).

The *Interim Policy* emphasizes that remedial personnel have a central role in making communities aware of these programs, and several interview participants reported that they routinely advertise them. Although none suggested that such initiatives might *directly* influence their decisions about how to remediate sites, some RPMs hinted that the politics of funding might play an indirect role in the decision-making process. One RPM, for example, maintained that the meaning of environmental justice at one of the sites he managed was simply "dollar signs":

> When a site gets tagged as an EJ site, that site starts getting money. When [the community surrounding his most controversial site] got wind of the EJ issue, they started "yelling and screaming." An EJ grant was issued to [a nearby university]; when community members got wind of that, they made a huge fuss ... Up until that time, EJ hadn't been an issue in the community. In my opinion, ever since that time, it's been about money.
> (Interview, RPM #1)

He denied that controversies over economic development had influenced his decisions in any way, but he suggested that they had empowered the community and helped it organize and publicize its cause. Another RPM highlighted the role of economic development at his most controversial site:

> He [a prominent community activist] and they [his organization] have had a larger idea all along. He's not just concerned about this site; he wants a complete redevelopment of the area, up to 500 acres. He has big ideas in mind, because the area is underserved in a market sense.
> (Interview, RPM #2)

Although this limited study can provide no evidence that EPA's economic redevelopment programs gave EJ communities leverage to secure more stringent remedies, it draws attention to the role of remedial personnel in delivering important neocommunitarian economic initiatives to marginalized communities.

Conclusion

The case study suggests that attempts to establish a scientifically verifiable "EJ community" failed in EPA Region 4. Clinton's EPA could not consolidate technocratic control over the boundaries of the EJ community, and it provided little clear guidance to street-level managers on how to make environmental justice an integral part of their decision-making. However, the Clinton administration's articulation of environmental justice subtly influenced both the distribution of hazardous waste risk and the trajectory of neoliberalization. While it did not lead remedial personnel to make decisions that might somehow redistribute risk more "justly," the Clinton EPA's emphasis on building trust and capacity in disadvantaged communities forced Region 4 to develop strategies for responding to allegations of discrimination in remedial decisions. Environmental justice policy affected the distribution of hazardous waste risk not by requiring more equitable distributions, but instead by making it necessary to secure community "buy-in" by responding to localized claims of environmental *in*justice. Also, by calling for remedial personnel to "enhance" community involvement for EJ communities and to advertise new opportunities to develop their economic and political self-sufficiency, Region 4's environmental justice policies transformed remedial personnel into agents for delivering neocommunitarian initiatives.

The case study raises the question of the degree to which environmental justice activism can form the basis for a politics of resistance to neoliberalization. As Peck and Tickell (2002) note, "deep neoliberalization" has opened up new points of vulnerability in neoliberal projects even while embedding them more deeply. Can environmental justice activism be a "basing point" for resistance, even after the Clinton administration's articulation of environmental justice as part of broader neocommunitarian initiatives? Or do the politics of environmental justice "revolve around axes the very essences of which have been neoliberalized" (Peck and Tickell 2002: 400)? In order to shed light on these questions, additional research should investigate how environmental justice activism and policy has matured and changed during the George W. Bush administration, itself a distinctive chapter in the history of neoliberalization.

Acknowledgments

A longer version of this chapter was Holifield, R. (2004) "Neoliberalism and Environmental Justice in the United States Environmental Protection Agency: Translating Policy into Managerial Practice in Hazardous Waste Remediation," *Geoforum* 35(3): 285–97. A previous version of this chapter was presented at the Association of American Geographers, Los Angeles, California, March 2002. The author would like to thank James McCarthy and Scott Prudham for organizing the AAG session and the *Geoforum* special issue, and Brooks Holifield, Steve Holloway, Hilda Kurtz, Deborah Martin, James McCarthy, Moira McDonald, Scott Prudham, Eric Sheppard, and Kristin Sziarto for providing insightful comments on earlier drafts. The author takes sole responsibility for the chapter's shortcomings.

This research was supported by a National Science Foundation Graduate Research Fellowship.

Notes

1 EPA requires that remedies reduce cancer risk to 10–14 to 10–16 and achieve a score of less than 1.0 on a non-cancer Hazard Index, but in many cases the site managers select a more stringent remedy (Viscusi and Hamilton 1999).
2 At the time this research was conducted, EPA had no national environmental justice policy and left it to its ten regional offices to develop their own interim policies.
3 Because of informants' concerns about confidentiality, most of my data consist of approximate quotations and paraphrases, with a limited number of sequences recorded as direct quotes. I present approximate quotes from the interview transcripts in italics, and I indicate sequences recorded as direct quotes by placing them in quotation marks.

References

Anderton, D. L., Oakes, J. M. and Egan, K. L. (1997) "Environmental Equity in Superfund: Demographics of the Discovery and Prioritization of Abandoned Toxic Sites," *Evaluation Review* 21: 3–26.

Brenner, N. and Theodore, N. (2002) "Cities and the Geographies of 'Actually Existing Neoliberalism,'" *Antipode* 34: 349–79.

Bullard, R. D. (1994) *Dumping in Dixie: Race, Class, and Environmental Quality*, Boulder, CO: Westview Press.

Clinton, W. J. (1994) "Executive Order 12898: On Federal Actions to Address Environmental Justice in Minority Populations and Low-Income Populations with Accompanying Memorandum," Washington, DC.

Cutter, S. L., Holm, D. and Clark, L. (1996) "The Role of Geographic Scale in Monitoring Environmental Justice," *Risk Analysis* 16: 517–26.

Foreman, C. H., Jr. (1998) *The Promise and Peril of Environmental Justice*, Washington, DC: Brookings Institution.

Gupta, S., Van Houtven, G. and Cropper, M. L. (1995) "Do Benefits and Costs Matter in Environmental Regulation? An Analysis of EPA Decisions under Superfund," in R. L. Revesz and R. B. Stewart (eds.) *Analyzing Superfund: Economics, Science, and Law*, Washington, DC: Resources for the Future.

Hird, J. A. (1994) *Superfund: The Political Economy of Environmental Risk*, Baltimore, MD: Johns Hopkins University Press.

Jessop, B. (2002) "Liberalism, Neoliberalism, and Urban Governance: A State-Theoretical Perspective," *Antipode* 34: 452–72.

Kurtz, H. (2002) "The Politics of Environmental Justice as the Politics of Scale: St. James Parish, Louisiana, and the Shintech Siting Controversy,", in A. Herod and M. W. Wright (eds.) *Geographies of Power: Placing Scale*, Oxford: Blackwell.

Lavelle, M. and Coyle, M. (1992) "Unequal Protection: The Racial Divide in Environmental Law," *The National Law Journal*, 21 September 1992: S1{-]S12.

Peck, J. and Tickell, A. (2002) "Neoliberalizing Space," *Antipode* 34: 380–404.

"Principles of Environmental Justice" (1991) Ratified at the First National People of Color Leadership Summit, October 1991, Washington, DC.

US Environmental Protection Agency Region 4 (1999) "Interim Policy to Identify and Address Potential Environmental Justice Areas," Atlanta, GA: Environmental Accountability Division.

Viscusi, W. K. and Hamilton, J. T. (1999) "Are Risk Regulators Rational? Evidence from Hazardous Waste Cleanup Decisions," *The American Economic Review*, 89: 1010–27.

Zimmerman, R. (1993) "Social Equity and Environmental Risk," *Risk Analysis* 13: 649–66.

Part III

Commentary

17 Neoliberal governmentalities

Wendy Larner

How should we understand contemporary changes in the governance of nature? Do these changes necessarily involve the imposition of marketised individualised processes that have negative consequences for nature? At first glance, the four chapters in this part would encourage us to be pessimistic. These analyses convincingly demonstrate the perils of neoliberalism: the eroding of public property through community gardening schemes in Vancouver; the failure to preserve public ownership of urban forests in Milwaukee; the targeting of disadvantaged communities in environmental justice programmes in the Southeastern United States; and the deaths in Walkerton following regulatory failure that led to the contamination of the town water supply. However, because they assume that new governmental forms are univocal and focus on their negative effects – described by the editors in their introduction as the weakening of environmental regulations, the loss of access for those who are less privileged, the introduction of marketised forms of governance – they only tell part of the story about the changing nature of environmental governance.

These articles examine key concepts in contemporary governmental processes, including property, privatisation, targeting and risk. There is a tendency to assume that these concepts are unequivocally associated with the political formation we have come to understand as neoliberalism. Yet reading across the four cases shows that each of these is more complex than initially expected. Indeed, seen together the chapters show that 'actually existing' environmental governance involves complex discursive formations, and that these discourses aren't always associated with those we might expect. For example, Nick Blomley's chapter suggests that community gardening schemes have complex genealogies that owe as much to the work of urbanist Jane Jacobs and subsequent efforts to reduce crime through environmental design, as they do with the individualised concepts of property associated with marketisation. Nik Heynen and Harold Perkins, albeit largely unintentionally, themselves reinscribe market logics when they use the discourses of the entrepreneurial city to support their ecological arguments for urban forests. In Ryan Holifield's discussion of the environmental justice movement, we see that the claims of grassroots activists, including those of community empowerment, citizen involvement and economic self-sufficiency, have had significant consequences for the Clinton administration's approach to the clean-up of hazardous waste sites. And even in the tragic case of Walkerton, we learn from Scott Prudham that the first steps towards environmental deregulation were not part of the 'Common Sense Revolution' of the conservative government of Mike Harris but rather were taken under the social democratic government of Bob Rae.

Why do the authors themselves not make more of these contradictions and inconsistencies? In large part this is because of the analytical position from which they start. While there are important differences in emphasis between them, they tend to assume that neoliberalism is a relatively coherent intellectual and political formation and that the analytical task is to show this formation is made manifest in the field of environmental governance. Whether the culprit is Newman's principles of defensible space, the continuation of Western imperialism under neoliberal capitalism, the Clinton administration, or the Common Sense Revolution in Ontario, the particular aspect of environmental governance explored in each chapter is seen as having a singular rationale. While Blomley subsequently reveals the narrowness of this approach by making visible the empirical complexities of property, and both Blomley and Holifield explicitly identify the multivocality of the political programmes with which they are concerned, ultimately all the authors remain tied to the task of recognising neoliberalism rather than taking the complexity of contemporary forms of environmental governance as their starting point.

Let me discuss each of the chapters in turn to demonstrate the implications this tendency has for their arguments, before concluding with some more general observations.

Blomley is interested in contemporary forms of public order regulation and policing. In this chapter he focuses on 'greenways', public gardening projects in which community use of a particular space is intended to send a signal to the 'disorderly' to go elsewhere. While greenways could be seen as the simple extension of private property rights over formerly public space, thereby confirming more general arguments about the neoliberalisation of space, this analysis usefully underlines the diversity of the concept of property. The assumption that public property is collective whereas private property is individualised not only obscures complex social histories, but also makes invisible a wide range of collective modes of 'private' property (communes, collectives, tribes, tenants in common, allotments, community gardens, rights of way, among others). Seen in this context, it is not simply 'ironic' that community gardens can be productive of citizenship and community. Moreover, not only do we 'need to unsettle core political categories, like property, acknowledging their diversity and loose ends', we also need to more carefully consider how concepts, strategies and techniques developed in one context come to be used in another.

Heynen and Perkins make the link between marketisation, the privatisation of property, and urban ecological crises in their analysis of the neoliberalisation of Milwaukee's urban forest. An interesting aspect of this chapter is the discussion of the changing rationales for tree planting over time. In the 1930s Depression, during which time Milwaukee's urban forests were significantly expanded, rationales for tree planting focused on the benefits of parks and playgrounds for the health and character of workers. In the present day, however, trees are to attract additional capital investment that would allow Milwaukee to successfully compete in the socially produced global hierarchy of cities. But why does increased interurban competition necessarily require reforestation? Why forests and not, for example, public art? Why plant more trees to reduce carbon dioxide rather than banning air conditioners? Why is the planting and maintenance of trees by the public sector the only viable means of providing green infrastructure? Following the example of greenways discussed earlier, could trees not be part of community gardening projects? In asking such questions I am not suggesting the authors are making unsubstantiated arguments, rather I am asking how claims about nature come to be framed in particular ways. This helps us understand why it is that the state, the city, an NGO, and indeed the authors themselves, find themselves caught up in market logics in their efforts to encourage tree planting in Milwaukee.

In his analysis of a US waste site remediation programme, Holifield argues that the Clinton administration's approach to environment justice involves a 'neocommunitarian strategy' that makes grassroots claims for greater empowerment and trust compatible with the processes of neoliberalisation. He shows how this hazardous waste programme constitutes 'environmental justice communities' by responding to localised claims of environmental injustice and then attempting to verify these claims through calculation, measurement and mapping. He concludes that rather than redistributing environmental risk more equitably, these efforts transform grassroots calls for public participation in environmental justice into technocratically managed claims of environmental injustice. His analysis tells us two very interesting things. First, it is clear that new forms of environmental governance do not simply ignore the claims of social movements. Consequently, the issue becomes that of determining when these governmental forms are indeed simply 'flanking projects' and when they represent real political gains. Second, and perhaps even more importantly, this case shows that efforts to consolidate technocratic control over governmental processes are not always successful, suggesting that neoliberalism can fail even on its own terms.

Finally, Prudham discusses the tragic case of Walkerton. Were the deaths from poisoned town water simply the result of administrative bungling or should they be attributed to something more sinister? Prudham's approach is to focus on the political economy of neoliberal regulatory regimes and show how they shape the increased probability that what he calls 'normal accidents' will occur. He argues that new environmental risks are being created because 'organized irresponsibility' is built into these regimes. Developing this claim through an analysis of measures promoted by the Harris government, including a regulatory milieu that privileged industry self-regulation, voluntary standards, and staff reductions, he shows that the deaths in Walkerton happened because of key gaps in regulatory oversight that were opened up or widened by neoliberal reforms. But are such regulatory gaps indeed an inevitable consequence of neoliberalism? Seen next to Holifield's chapter, it becomes clear that the Common Sense Revolution involved particularly raw forms of environmental governance in which market rhetoric did indeed become regulatory failure. But it is important to remember that neoliberal techniques can be and often are, articulated to different political formations. Environmental standards, for example, can be voluntary or mandatory. Privatisation and deregulation do not necessarily go together. Audit processes can involve regular reporting. Seen in this context, our discussions of neoliberalism need to delineate between the absence of regulation and new regulatory forms, and explore in more detail the different ways in which the latter are used and the effects they have.

It may seem that I am being overly critical of four strong and compelling chapters which reveal much about the inadequacies of contemporary approaches to environmental governance. This is not my intention. What I am suggesting, however, is that if we do indeed live in an era in which neoliberalism has been normalised, we need to take seriously the complexity of real examples. Not only will this help us to think about different forms of environmental governance, it is also important if we are to discern more or less progressive political possibilities. To return to the chapters, the approach to environmental governance promoted by the Common Sense Revolution is clearly highly problematic; greenways are apparently much less so. The task of neoliberals has been to present their unwieldy and contradictory political assemblage as a coherent ideological and institutional formation with necessary outcomes. Are we, as critical scholars, also unintentionally contributing to these ambitions?

18 Neoliberal environments, technologies of governance and governance of technologies

Dianne Rocheleau

As I read the four chapters in this part I was transported back to a news story in May of 1996 about the crash of the ValuJet airliner in the Everglades just outside Miami. It occurred to me in retrospect that this was a watershed moment, a public debut of neoliberal governance and disasters-by-default through the kind of stealth planning that Prudham describes – in Günter Grass' words – as "planned irresponsibility". The proximate cause of the ValuJet crash, and the deaths of 110 people, was an explosion of illegal hazardous cargo (used and partially full oxygen canisters) in a passenger plane. The subsequent investigation and discussions of accountability and blame revolved around the practice and alibi of "out-sourcing".

A statement on the website posted by the victim's families notes that

> It is an incontrovertible fact that ValuJet ordered SabreTech [a maintenance subcontractor] to return the expired oxygen generators to them rather than … dispose of [them] as hazardous waste, and … knowingly transported them on board Flight 592.

ValuJet (now merged into AirTran) later accused SabreTech of disguising and loading the generators on Flight 592 "against the airline's wishes". Yet, statements by ValuJet directly after the crash reflected their awareness of the oxygen generators on board the flight. The families asked: "[W]hat possible reason could ValuJet have … for transporting worthless, expired oxygen generators to Atlanta?" The answer was as simple as it was morally stunning. Prior to this crash, airlines routinely transported oxygen generators back to their home bases to dispose of them at lower cost. ValuJet was not authorized (apparently with good reason) to carry hazardous material. The final conclusion of the victim's families could apply to any number of environmental, water supply, food supply and transportation disasters. "We must live the rest of our lives with the sickening realization that our loved ones are dead because ValuJet Airlines was attempting to save a few dollars in hazardous waste disposal costs."

Fast forward back to 2006 and, at least in the U.S., we are bound to take for granted all manner of out-sourcing as a matter of course, even in that last and most sacred bastion of government prerogatives, the making of war. Names like Blackwater Security and Halliburton come to mind. Everything from the provision of food, weapons and safety equipment (or not) and the conduct of misconduct, reportedly including torture, assassination and shoddy hazardous waste disposal, has been outsourced to companies whose boards and employees are not subject to the same terms of accountability as soldiers or civil servants.

The concept of governmentality provides us with some key tools to make sense of this phenomenon with respect to the environment in general and the specific cases presented in the four chapters. In the cases I noted above, the politics of plausible deniability through out-sourcing are a very explicit, purposely designed variant of a broader class of technologies of governance employed under neoliberalism. As Prudham notes, there are specific neoliberalizations, not just an amorphous ambience of neoliberalism. These may take the form of deregulation, decentralization, devolution, privatization, perverse versions of public participation, and outsourcing, each of which occurs in at least one case study in this section.

Raymond Bryant's work, *The Conduct of Misconduct* (2002) on the regulation of conservation and biodiversity in the Philippines provides a window on the way that apparently progressive government institutions, non-government organizations (NGOs) and other private or semi-private actors can become part of state practice, internalizing agendas in ways that they themselves, their members or their constituencies, may not predict or intend. Under this approach, the Trojan horse does not breach the city gates but enters the imagination, re-structures the sense of self, and re-directs the practices of willing subjects (see Agrawal 2005, on Environmentality).

This is not a one-way process, however. For example, the politics and rhetoric of decentralization are widely recognized in the developing world as a mixed bag, often simply a guise for re-structured, remote control "indirect rule", long a ploy of Colonial Empires, but also open to manipulation from below. Ask the elder Akamba women who did forced conservation labor in Kenya under the Colonial government during World War II, and later worked on soil and water projects for the promise of patronage and famine relief under Moi's government in the 1980s (Thomas-Slayter and Rocheleau 1995). Forced, coerced and/or cajoled labor was conducted through pre-existing communal self-help organizations. In the communal work title, however, the substantial community labor and the full support of the state were often counter-appropriated to accomplish evolving political, agricultural and environmental purposes of participating women. This *counter-assimilation* is part of an under-theorized and overlooked governmentality and environmentality from below, comprised of technologies of resistance, solidarity and self-governance under diverse conditions.

The governance literature and the four cases in this part feature some treatment of both technocratic and grassroots self-regulating environmentalists constructed as Neo-Liberal subjects. Here, the new Neoliberal subject conflates self-interest with the interest rates on savings, loans and credit cards and rates of taxation. The self merges with consumer, owner, and sales agent. The counter-appropriation of new identities and the re-constitution of Environmentalism and Environmental institutions from below and between, however, are somewhat under-theorized. It is equally imperative to investigate new types of Neo-liberated subjects (protagonists of Fair Trade, Diverse Economies, Autonomous Communities, and a resurgence of the Commons) who may subvert the very basis of conservative neoliberal politics through new kinds of subjects, commerce and models of semi-autonomous local governance.

Lessons from the four cases studies

The contribution of this volume and these four articles is to link the technologies of environmental governance with the "progress" and processes of Neoliberalism, and specific Neolibrealizations, as noted by Prudham. Each of the case studies illustrates an example of either trying to involve people in responding to what should have been preventable damage, in preventing future damage, or trying to garner public complicity in the avoidance of responsibility for predictable and avoidable damage or even disaster (before and after the fact). The rise of technologies of resistance, solidarity and/or self-governance from below, under Neoliberalism, is also present, though generally submerged in all of the cases except Blomley.

Walkerton, Ontario

When I read Prudham's account of the Walkerton, Ontario, case I find myself halfway between the ValuJet crash and the war crimes of the U.S. wars in Afghanistan and Iraq. The Walkerton story exemplifies the explicit use of plausible deniability, as well as the more indirect technologies of progressive, multiple and intersecting instances of deregulation and out-sourcing. The practice of out-sourcing was very much at the center of Prudham's story about the poisoning of the water supply on Victoria Day weekend of 2000. The sub-contracting of dangerous, dubious or potentially expensive and difficult jobs to contractors whose standards are known to violate government rules and standards is a way to avoid both compliance with the rules and subsequent accountability for misconduct. The deaths of people in Walkerton as well as in the ValuJet crash occurred in a twilight zone between planned irresponsibility, by default, and the carefully designed defense of plausible deniability. The involvement of a third party in the pre-programmed "conduct of misconduct" (Bryant 2002) allows for a slippage in blame and accountability. In this case we also see the sleight of hand shift, from accountability to accounting, and part way back again, as the measure of the state's duty and responsibility in the provision of clean water. The first shift is accomplished through several incremental layers of seemingly benign tax cuts and re-engineering of banal government functions, a familiar practice of the retreating state which is actually more actively involved in the sale and betrayal of the public trust than retreat *per se*.

Strathcona, Vancouver

In his article on the urban gardens of the Strathcona neighborhood of Vancouver, Blomley proposes the concept of property as a complex construct with radical potential, and as such presents the strongest case for counter-assimilation, technologies of self-governance and to some extent solidarity. The gardens are publicly presented as part of a move to privatize the streets and adjoining green spaces to exclude and/or police the presence of people deemed dangerous, or undesirable, by at least some state actors and neighborhood residents. Yet the process and practice of gardening creates a more inclusive and collective space with the resources of both the state and participating residents. The surprise and the counter-moves, planned or not, re-assert a place for common property, or more correctly, a neighborhood commons on public land, managed and maintained by an assemblage of private individuals who become a community partly through the practice of making the gardens. This is an element of governmentality that is not as developed in the theoretical works cited above or in the other case studies.

Bromley's call for us to address common property is well taken but we should engage the diverse interdisciplinary work underway on this theme since 1985. His case, and his excellent conclusion, would be far more grounded and convincing if it were also to connect with the Community Gardens literature, as well as the body of work on Common Property Resources (CPR) theory and management, than tie it all together through Gibson-Graham's (1996) theoretical analysis. While much of the CPR work has focused on fairly economistic and legal definitions of common property, there is also a rich literature on the Commons, beyond property *per se*. The land struggle movements of the last three decades, from rural forest and farm communities in Brazil (Wolford 2006) to the urban gardens of New York City, have been waged through collective action of various kinds, to secure or protect property as the basis for community life. These embody the radical potential and the limitations of property as both a tool of power and a tool of resistance, autonomy and self-governance.

EPA Superfund sites in the Southeastern US

Holifield's article, like Bromley's, complicates the concept of governmentality, though in different and significant ways. In a section that hearkens back to the points raised in Bryant's article on governmentality and biodiversity in the Philippines, Holifield reveals and analyzes the contradictory views within the responsible state agencies as to what constitutes proper conduct. The various interpretations and implementation of the Clinton Administration Executive Order No. 12898 ("Federal Actions to Address Environmental Justice (EJ) in Minority Populations and Low-Income Populations") illustrates this conflict, and embody the kind of slippage that Bromley also mentions in a more generic sense in his conclusion.

Holifield's quotes from the various technical officers and scientists in the US Environmental Protection Agency's (EPA) Superfund program demonstrate the different approaches to the mandate to identify and provide extra resources and attention to EJ communities. Some technical and scientific personnel saw this presidential directive as an opportunity to mobilize and pro-actively serve "minority" and low-income communities. Others saw it as "unfair" or "lacking in scientific objectivity" and reluctantly accepted it as a newly mandated constraint. However, Holifield shows that the mere fact of its existence provided certain leverage to communities organized by motivated technical personnel, or from within, to secure additional legal and technical services as EJ communities. The naming of this phenomenon and the recognition of racism as a factor in environmental risks, hazards, damage, and clean-up, seems to have added an element of self-oversight among the technical staff. The elusive nature of the EJ phenomenon proved impossible to quantify and label "scientifically" from outside and left plenty of slippage for active and creative ways to use this explicit recognition of an ambiguously defined condition. Finally, Holifield provides strong evidence that we are dealing with more than economic competition in terms of rational economic self interest. The selective dumping of toxics and the skewed choice of sites for subsequent remediation is about race and class privilege, and culturally ingrained constructs of vulnerability, entitlement and rationalization of irrational inequalities. This case provides some of the best evidence for the cultural preconditions of neoliberalism.

Milwaukee's disappearing urban forest

Heynen and Perkins provide an account of the making, abandonment and selective reconstruction of the urban forest in Milwaukee, Wisconsin, from its creation by a socialist city government and populace in the early twentieth century to its current decline and selective rehabilitation under Neoliberal terms of governance. The partial reconstruction of the urban forest is being conducted by technocratic, managerial agents of government operating at a distance through the convergent visions, values, labor and financial investment of middle and upper class homeowners. The retreat and selective return of the state in the management of the urban forest are couched in terms of property values as well as local and larger environmental services. Heynan and Perkins present the local state's role in the making and abandonment of the urban forest, and treat the material environmental changes and ecological processes, and their significance at local, regional and global level, as well as their entanglement with property, class and race.

However, an analysis of Neoliberalism and environment is not complete without an equally thorough treatment of people's own experiences, analysis and values, as well as their actions and initiatives. Who plants trees and why? Who wants or doesn't want trees, where, what kind, and why? Are there urban gardens or other community-based initiatives to plant or care for trees, other kinds of greenery or common space? If people in systematically neglected neighborhoods in Milwaukee do want trees or some kind of green spaces, are they planting any urban gardens, with or without trees?

People evaluate and value trees and forests very differently from one place to another and from one subject position to another. Alec Brownlow's (2006) work in Fairmont Park in Philadelphia suggests that some kinds of trees and green space may become hazards to the safe passage through areas already plagued by high rates of violent crime. In contrast, middle- and upper-class homeowners in Cambridge, Massachusetts, were busy in 2003 actively deforesting some neighborhoods as part of a gentrification process, seeking to let in more light, provide a view from new picture win- dows, show off re-designed building fronts and signal affluence, all at the expense of 50–100-year-old trees. Assuming that street trees or the urban forest *per se* are universally desired features promoted and owned by the rich potentially overlooks the complex relationships between neoliberal technologies of governance, environmental quality and (in)justice.

Conclusion

What is demonstrated in these four case studies, and what can we learn from them? I challenge the authors in this part, and all of us concerned with the phenomenon of neoliberal regimes and technologies of environmental governance, to turn our attention to the following five points:

1 conflicting notions of appropriate conduct within and between state and corporate spheres, as both indicators of complexity and opportunities for alternatives;
2 persistence and resurgence of "the Complex Commons" threaded over, under and through various property regimes (state, private and common);
3 making and re-making of environmental subjects and socio-natures by cultures (political, economic and otherwise);
4 cultural preconditions for Neoliberal technologies of environmental gov- ernance; and
5 technologies of governance from below and between, including technol- ogies of resistance, self-governance and solidarity.

References

Agrawal, A. (2005) *Environmentality: Technologies of Government in the Making of Subjects*, Durham, NC: Duke University Press.

Brownlow, A. (2006) "An Archeology of Fear and Environmental Change in Philadelphia", *Geoforum.* 37: 227–45.

Bryant, R. L. (2002) "Non-governmental Organizations and Governmentality: 'Consuming' Biodiversity and Indigenous People in the Philippines," *Political Studies* 50(2): 268–92.

Escobar, A. (1994) *Encountering Development: The Making and Unmaking of the Third World*, Princeton, NJ: Princeton University Press.

Gibson-Graham J. K. (1996) *The End of Capitalism (As We Know It): A Feminist Critique of Political Economy*, Oxford: Blackwell.

Thomas-Slayter, B. and Rocheleau, D. (1995) *Gender, Environment and Development in Kenya*, Boulder, CO: Lynn Reinner.

Wolford, W. (in press) *This Land is Ours Now: Social Mobilization and Sugarcane in the Brazilian Northeast*, Durham, NC: Duke University Press.

Part IV

Resistance

19 A "continuous and ample supply"

Sustained yield timber production in northern New Mexico

David Correia

In 1948, the U.S. Forest Service introduced sweeping grazing reductions in the El Rito Ranger District in the Carson National Forest in northern New Mexico. The restrictions transformed a subsistence economy organized around irrigated agriculture and the communal use of upland rangelands and forests into a wage-dependent economy based on commercial forestry. The change was based on a 1947 case study in which local forest rangers claimed that smallholder Hispanic sheepherders "caused surrounding national forest ranges to become depleted of vegetative cover to such an extent that a reduction in permitted grazing use is necessary."[1] The district proposed to remedy the economic hardship anticipated by the "stock reduction program" by increasing timber-related jobs through the creation of a special sustained yield timber production unit.

Based on the 1944 Sustained Yield Forest Management Act (SYFMA), the sustained yield unit promised

> [to] promote the stability of forest industries, of employment, of communities, and of taxable forest wealth, through continuous and ample supply of forest products: and in order to secure the benefits of forests in maintenance of water supply, regulation of stream flow, prevention of soil erosion, amelioration of climate, and preservation of wildlife.[2]

This chapter examines U.S. Forest Service efforts to regulate nature and labor in service to industrial timber production in northern New Mexico.[3] Rather than benefiting the local economy and local ecology as promised in the language of SYFMA, I argue that sustained yield policies produced the conditions necessary for capital to overcome local common-property resource practices, reserving timber resources and local labor power for industrial interests. The Forest Service accomplished this transformation by mediating capital's access to nature through a state–private timber monopoly. Access to cheap labor was guaranteed through the application of labor standards that placed locals in a dependent relationship to outside commercial interests. Access to timber was guaranteed by restricting competition, removing economic barriers to production, and defining sustained yield solely in economic terms.

Regulation and accumulation in forestry: an ecological marxist framework for analysis

The commodification of nature – from human organs (Scheper-Hughes 2000), to water (Swyngedouw 1997), to whole ecosystems (Robertson 2000) – produces contradictions within environmental systems and capitalist production itself that imperil the socio-economic and biophysical foundation of human and non-human life. Among the contradictions of capitalist production includes the fiction of treating human labor power and natural resources as commodities. Yet, labor power and natural resources are not commodities at all – not "things" produced in the image of capital – but rather exist separate and outside this process and cannot be produced capitalistically. Unfettered access to nature and labor requires the intervention of the state, yet this arrangement provokes accumulation crises as the very natures upon which resource-dependent industries rely are treated as exhaustible commodities. In addition, regulatory measures designed to protect land, labor and money from exhaustion invariably "impaired the self-regulation of the market, disorganized industrial life, and thus endangered society in yet another way" (Polanyi [1944] 1957: 3–4). This contradiction provides a point of departure for an ecological Marxist theory of capital accumulation that recognizes a specific internal contradiction in capitalist uses of nature. Since labor power and nature are not commodities produced in the metabolism of capitalist production, *they are only available as commodities through the intervention of the state*:

> This means that whether or not raw materials and needed labor skills and useful spatial and infrastructural configurations are available to capital in requisite quantities and qualities and at the right times and places depends on the political power of capital, the power of social movements that challenge particular capitalist forms of production conditions (e.g., struggles over land as means of production versus means of consumption), state structures that mediate or screen struggles over the definition and use of production conditions (e.g., zoning boards), and so on.
>
> (O'Connor 1998: 165)

This chapter explores the contradictions and consequences of regulatory efforts in New Mexico that sought to counter the problems of accumulation in the resource-dependent forestry sector. Cheap access to land and labor allowed the industry to operate in an unstable pattern that threatened the continued accumulation of capital. Such practices resulted in cycles of overproduction creating price fluctuations that then led to underproduction, layoffs, and mill closures. Further, the resulting itinerant nature of the industry fostered a floating reserve army of workers as employment levels constantly shifted across space and time (Cronon 1991; Clary 1986). Throughout the early twentieth century, timber industry leaders were unable to work out these contradictions through market mechanisms or cartel agreements.

In the 1920s, the timber industry's lobbying organization, the Western Forestry and Conservation Association, began to push for sustained yield legislation as a means to stabilize corporate profits for the largest operators. As Robbins has argued,

> The essential function of those industrial proposals, however, was never the humanitarian end of community stability; rather they would serve to restrain production and compel more rational market behavior. The forest in that sense became a commodity to be manipulated by private enterprise for private ends.
>
> (1987: 190)

Sustained yield was a political dream of West Coast lumber barons – a means to secure access to forests in the Pacific Northwest. As this chapter documents, sustained yield, a means to resolve the cycles of crises in industrial timber production in the Pacific Northwest, was first implemented in northern New Mexico.

The Vallecitos Federal Sustained Yield Unit

In a public meeting held in Vallecitos in 1947 to consider creating the sustained yield unit, one resident stood up and bitterly condemned the Forest Service for their treatment of local Hispanic shepherds. He drew enthusiastic applause from his neighbors by comparing Forest Service policies as equivalent to "[putting] a yoke around our necks."[4] Despite internal Forest Service memoranda linking the Unit to a comprehensive round of grazing restrictions on local federal lands, the forest supervisor replied that "this was a question aside from the purpose of the meeting."[5]

The suspicions articulated by local residents in that meeting stem largely from a history of land and water expropriation that began in 1846 with the invasion of Mexico by United States troops. In 1848, the signing of the Treaty of Guadalupe Hidalgo officially ended the Mexican-American War and ceded 50 percent of Mexico's territory, including the Territory of New Mexico, to the United States. Throughout the periods of Spanish and Mexican control of the region, grants of land to individuals and agricultural communities served as a means to promote settlement in the region. The Treaty of Guadalupe Hidalgo established that these grants would be "inviolably respected" by the United States (Ellis 1975: 14).

The adjudication of Spanish and Mexican land grants in the territory proved onerous for land grantees, however, as only 155 of the 295 New Mexican land grants conveyed during Spanish and Mexican control of the region were confirmed, accounting for only 2 million of the nearly 35 million acres granted (Ebright 1994; Poling and Kasdan 2001). The poor record of confirmation partly stems from an 1897 Supreme Court case, *U.S. v. Sandoval* (67 U.S. 278, 1897) in which the Supreme Court found that

Mexico had granted only *possession*, not *title*, to the community grants made between 1821 and 1848.[6] As a result of this finding, under the terms of the Treaty of Guadalupe Hidalgo, the United States, not land grantees, received ownership of most of the community land grant commons.

Among those grants affected by the *Sandoval* case were the Vallecito de Lovato and the Petaca Grant, both rejected by the U.S. Supreme Court.[7] All of the Vallecitos Federal Sustained Yield Unit is made up of land originally claimed by the Town of Vallecito de Lovato and the Petaca land grants. In that initial public meeting held in 1947, local residents feared that the operation and management of the VFSYU would further curtail access to forest resources.

On January 21, 1948, the Vallecitos Federal Sustained Yield Unit was declared active. Prior to the Unit, timber operations cut over 7 million board feet (mmbf) per year. In establishing the VFSYU, foresters set the sustained annual yield at 1.5 mmbf.[8]

On March 17, 1948, the Vallecitos Lumber Company, an inexperienced outside outfit recruited by Carson rangers self-described, rather unimpressively, as having "at one time or another, during [our] respective careers, done some logging," requested the designation of Approved Responsible Operator (ARO), a category that conferred on them oligopsony status for all timber sales on the Unit.[9] In return for this advantage, the ARO was to send at least 40 percent of all harvested timber to the local mill and maintain a workforce where, under the original Forest Service language, "not less than 90 percent of labor employed by Vallecitos shall be local residents."[10] In their initial letter, however, the Vallecitos Lumber Company agreed "to employ, *in other than supervisory position*, at least 90 percent local labor" (emphasis added).[11] Two weeks later, the Forest Service issued a special clause appended to VFSYU sale agreements wholly adopting this language.

The policy change was a significant one for the future of the Unit in Vallecitos. In the immediate future, however, it wouldn't matter. The president of the Vallecitos Lumber Company disappeared before ever entering into a sale agreement with the Forest Service. Until another operator, the Jackson Lumber Company, was named ARO in October of 1952, the VFSYU was idle; grazing restrictions, however, remained in place.

In 1952, the district responded to requests from Jackson Lumber and more than doubled the annual sustained yield from 1.5 mmbf to 3.5 mmbf.[12] Within two months of Jackson's tenure, complaints began to surface regarding hiring practices. Local residents frequently complained to the Forest Service, United States Senator Dennis Chavez, and local elected officials about Jackson's unfair labor practices. Jackson repeatedly asked for exemptions from the 90 percent labor standards, once arguing that "*competent* men are not available locally" (underline original).[13] The Forest Service often allowed exemption requests to the labor standards, arguing that it just "took time to get the local people accustomed to the regular routine of going to work at an industrial plant every working day."[14] Frequently Jackson requested exemptions for positions they described as "untrainable specialists,"[15] despite the fact that the employment office in Española held applications by experienced local workers for just these positions.[16] By 1955, even the district ranger began to believe Jackson "rigged" the labor lists.[17]

Local workers pursued individual and collective strategies in response to company labor violations. While individual workers solicited the Forest Service to enforce labor standards, woods and mill workers also began collectively to organize in the summer of 1955. Jackson, and the timber firms that followed, retaliated against individual worker complaints with a pattern of intimidation that sought to discredit worker protests with the Forest Service. In addition, collective challenges by workers were countered with harsh tactics designed to atomize workers. For example, immediately following public efforts by workers to organize in 1955, Jackson fired 18 workers it suspected of union affiliation.[18] In July of 1955, despite Jackson's union busting tactics, local workers succeeded in organizing the Lumber & Sawmill Workers Local Union No. 2507. On the 20th of July, the locals went out on strike. Despite the strike, according to VFSYU labor regulations, the ARO was still required to meet local hiring standards. Despite this advantage, the union was unable to either shut down the operation or force Jackson into negotiations. Instead, the company circumvented labor requirements by purchasing private timber sales and "mixing" the lumber, which they argued allowed them to employ a larger percentage of non-local men. While officially opposing the practice, the district did little to restrict Jackson. In 1957, the district forester in El Rito admitted: "timber inspectors from the Washington Office have expressed some amasement [*sic*] when they learned our purchaser was permitted to bid competitively outside the unit and haul the timber into said unit."[19]

Against charges of improper firings made before and during the strike, Jackson maintained that all the fired workers were "habitual drinkers." Jackson attempted to deflect attention from their record of labor standard violations by challenging the authority of local labor leaders. Despite frequently hiring workers from outside the employment area, Jackson claimed in a letter to the forest supervisor that the president of the union lived in Española and should therefore be barred from employment and union activity on the VFSYU.[20] Despite Jackson's constant attempts to evade the 90 percent local labor requirement, it remained an issue the company could not avoid during the strike. By dragging out negotiations and forestalling Forest Service retaliation, Jackson continued production in both the woods and mill operations throughout the summer and fall of 1955.

Eventually local workers forced the Forest Service to hold Jackson accountable. In November 1955, a hearing in Santa Fe allowed the union to confront Jackson on documented labor violations. During the hearing, the company's lawyer complained:

If we are now told that even though the individuals are available, and that we have to go to them, and have to employ them on their terms, then we are being told that we have to enter into a contract with the union specifying wage rates and hours of work that the union dictates; in other words, we have no free choice, and the men themselves have no free choice.[21]

The primary issue in the hearing was the use of outside labor and discrepancies in wage scales, all violations of Unit policies. In addition, Jackson Lumber sought to move pay schedules from an hourly wage scale to piece-wage work in an effort to intensify production while lowering costs. In a hearing discussing the labor conflicts, Jackson's attorney defended both the use of non-local labor and the piece wage schedule:

> these lumber stackers stacked on a top truck basis where they cruised timber we were getting from private lands and roughly at $1.00 a thousand they made about twice as much wage per day as the [local] lumber stackers working by the hour, so after that the lumber stackers that had refused to stack by the thousand then went on the thousand rate and they jumped their wages from $8.40 a day to where they were making $12.00 and $15.00 a day, and Jackson's lumber that had formerly cost him $1.50 a thousand to stack was getting stacked for a $1.00 a thousand, so both the local people and the company benefited from the importation of the lumber stackers.[22]

The union's lawyer was not convinced:

> Mr. Jackson is one of very few people in this country who is fortunate to have a sustained yield unit contract, and he is enjoying privileges under that contract that no private operator enjoys and here he is bringing in this foreign labor from Texas to work in New Mexico in his mill to save 50 cents on the thousand in piling lumber. I think that's very absurd.[23]

By restricting locals to positions as laborers, and regularly offering exemptions to the 90 percent local-labor rule, the Forest Service facilitated the enrollment of locals into a wage economy as a reserve army of workers. Local resistance to this process, however, was fierce. Striking workers actively picketed the mill during the strike, harassing Jackson's managers and scab workers imported from outside New Mexico. Forest Service intransigence to union pressure, and its refusal to act quickly in the summer of 1955, doomed the strike. Had the Forest Service enforced the labor standards, as required, the strike would have shut down all timber operations. The strike and the November hearing changed little for workers regarding wages and job security. Following the hearing, logging workers had two options: return to work without the protection of a union contract or see the VFSYU dissolve – in which case Jackson would hire only non-local labor. After a nearly two-year fight between the union and Jackson lumber, the conflict ended with a fire in 1957 that destroyed the sawmill. Despite intense investigation of what was clearly arson, no arrests were made. The company asked the Forest Service to end its designation as approved responsible operator.

The Unit was without an operator for 15 years. Throughout that time, the importance of the lumber and wood products industry in Rio Arriba declined. By 1958, the U.S. Forest Service began to plan for further grazing reductions and the decommissioning of the VFSYU.

By the middle of the 1960s, however, the district began to feel pressure from land grant activists, led by Reies Lopez Tijerina and his organization, *La Alianza Federal de Mercedes*. La Alianza was a state-wide organization of land-grant heirs organized to advocate for the return of all Spanish and Mexican land grants. In a confidential plan approved on February 9, 1967, for the abandonment of the VFSYU, the rangers prepared a strategy to slowly convince local people to support ending the Unit. In the plan district rangers recognized their precarious position:

> It is expected that the Federal Alliance of Land Grants (Alianza federal de las Mercedes) will see this as an opportunity to move in and try to stir up support for their program. We will attempt to keep the proposal for abandonment of the sustained yield unit separate from the grazing problem, which is now under consideration. However, from past experience, it has been shown that the local people will be inclined to associate the two.[24]

Plans for abandonment, however, changed in late spring. On June 5, 1967, armed men from *La Alianza* raided the Rio Arriba County Courthouse in Tierra Amarilla in an act they hoped would illustrate the plight of land grant heirs in New Mexico. Following a long history of often violent resistance by Hispanos to the land and water expropriation that followed the arrival of the United States (see Rosenbaum and Larson 1987), the courthouse raid, and the subsequent land grant war – in which Governor David Cargo ordered National Guard tanks into the mountains to hunt down *Alianza* raiders – brought national media attention to the social and economic conditions in Northern New Mexico (Gardner 1971). The "TA Raiders," as they came to be known, received widespread popular support, particularly among the residents of Vallecitos, Cañon Plaza and Petaca, most of whom were land grant heirs.

Despite the raid, Yale Weinstein, the president of Albuquerque-based logging company, Duke City Lumber, sought to resuscitate plans to abandon the Unit. Weinstein saw a terminated VFSYU as a way to bust the union at his nearby Española mill by using workers from Vallecitos and the surrounding area as scabs: "We have been involved in a labor dispute. Our Company is unfortunately caught between two unions attempting to achieve recognition, and we are presently facing several unfair labor charges filed with the NLRB."[25] In a letter to the El Rito district office, Weinstein proposed an end to the VFSYU: "Assuming that we could successfully terminate the VFSYU, we would be very agreeable to making a long-term logging contract with Mr. Abiniano Gurule or any other local people from Vallecitos of Cañon Plaza."[26]

Carson Regional Forester William Hurst, however, directed local staff to stall decommissioning plans:

> the situation in northern New Mexico is now receiving national attention. It would, in my opinion, be wise to continue the Vallecitos Federal Sustained Yield Unit ... if we propose the elimination, many people will automatically be against it. It will also appear that we are taking something more away from the "poor people of Rio Arriba County."[27]

The Forest Service developed and publicized plans to support small local industries in support of local economic development. But Forest Service rhetoric to help locals proved hollow, as requests to name a local, cooperative timber outfit as ARO were denied by the Forest Service. In 1972, the Forest Service designated Yale Weinstein's Duke City Lumber as approved operator.

Duke, much like Jackson Lumber before it, routinely violated labor agreements and contractual obligations. In November 1970, they hired only 44 percent local labor. In December 1970, they increased it to 50 percent local labor.[28] Local residents frequently complained to the Forest Service about Duke City's labor practices.[29]

Local foresters, however, emphasized to workers "the mutual responsibility" they shared with the company in the operation of the mill and work in the woods. "The people have a responsibility to come to work on time and on a regular basis," a district forester said.[30] A local labor leader named Albert Jaramillo argued that "local people left the woods because of late payment, and in some cases non-payment of wages."[31]

By February 19, 1971, Duke City still maintained only 47 percent local labor in their timber operation.[32] Duke City continued to argue that absences and poor work performance attributed to the low percentages. Local residents, however, claimed that local men were made to look bad as a means to avoid hiring them. Duke City's subcontractors gave out faulty equipment and then fired local workers when they couldn't maintain production quotas.[33]

In 1974, Jaramillo wrote to Yale Weinstein complaining about Duke's subcontractor: "Do we have to be Bill Thomas's slaves?" asked Jaramillo. He cited a litany of complaints, including unfair hiring practices, wage scales favoring non-local labor, and overtime paid to outsiders but withheld to local workers. "Since you told me to contact you if we had any trouble up here I think this is one of the greatest problems we can ever encounter. I would appreciate any help you can give us," he concluded.[34] Weinstein forwarded the letter to the Carson Forest Supervisor saying, "[t]he problem of working with the people in Vallecitos has always been a rather sensitive one and I will work with Bill Thomas and Dave Halder to try to keep the labor situation as 'cool' as possible."[35] In the same letter, Weinstein requested a reduction in stumpage costs to make up for what he viewed as the costs of additional security to protect his subcontractors against the threat of violence he inferred from Jaramillo.

In the midst of Duke City labor violations and disgruntled local employees, the district increased the allowable annual cut offered to Duke City from 3.5 mmbf, where it had been since 1952, to 4.0 mmbf. In June 1977, the Vallecitos Sawmill was destroyed by fire.[36] Despite the lack of evidence indicating arson, a number of acts of vandalism and arson did occur in the late 1970s, including repeated acts of vandalism to the equipment of Duke's subcontractor Bill Thomas. Thomas gave up his contracts in the Unit, while Duke rebuilt the mill and continued to operate in Vallecitos until 1996.

Throughout this period, Duke City continued to argue that the allowable annual cut was too small. In 1980 the annual cut was increased from 4.0 to 4.2 mmbf. In 1985, the draft forest plan proposed increasing the cut to 8.0 mmbf. The furor that erupted over the plan, led by local residents who argued that the plan exceeded the sustained yield of the area and represented a gift to a company that had made it a practice to violate labor standards, resulted in a negotiated settlement reducing the proposed annual cut. Duke City received 5.5 mmbf, with an additional 1.0 mmbf of timber and 1.1 mmbf of small forest products reserved for locally-owned operators.

The activism of local residents and loggers forced Duke City to replace outside contractors and employ two local timber operations. Unfortunately, the successes of the local operators were limited by constant conflict with both Duke City Lumber and the Forest Service (Krahl and Henderson 1998; Wilmsen 2001). In addition, during the 1990s, lawsuits involving the Forest Service, local environmental organizations seeking protection for endangered species, and local loggers arguing for increased local control of resources further curtailed timber-related jobs in the area.

Conclusion

This chapter considered the history of the Vallecitos Federal Sustained Yield Unit not as a narrow example of capital's exploitation of nature, but rather as a broader examination of the social context of production and regulation of nature. In the case study, sustained yield, designed to resolve the contradictions of capital's use of nature in Pacific Northwest forests, was incapable of resolving the contradictions that emerged based on the specific historical, geographical and ecological conditions of production in New Mexico. The Unit guaranteed timber and an available workforce to commercial timber operators under the rationale that this radical restructuring of the local economy would produce benefits that would somehow trickle down to local people. The policies and practices of sustained yield forestry exacerbated, rather than resolved, the contradictions of industrial forestry.

Forest Service practices in regulating access to nature and labor along with the practices of the timber operators in manipulating labor costs through producing a reserve army of workers were designed to establish conditions of production favorable to industrial timber interests. These efforts were only partly effective. The practices and institutions of sustained yield forestry opened a space for local resistance to capital. The suspicions of local shepherds to the practices and discourses of federal bureaucracies stemmed from a long history of land and water expropriation in the region. The operation and administration of the Unit, however, failed to resolve the contradictions of resource use, solidified local resistance to Forest Service policies, and impaired the operation of the very markets in nature and labor the regulatory framework sought to establish.

Acknowledgment

A longer version of this chapter was Correia, D. (2005) "From Agropastoralism to Sustained Yield Forestry: Industrial Restructuring, Rural Change, and the Land-grant Vommons in Northern New Mexico," *Capitalism, Nature, Socialism* 16(1): 25–44.

Notes

1 27 March 1947 Case Study. VFSYU 1: 3A, 21.
2 SYFMA, March 29, 1944 (58 Stat 132).
3 This chapter is based on the examination of Forest Service archival documents at the Carson National Forest office in El Rito, New Mexico conducted from 2004–2005.
4 11 December 1947 Public meeting report filed with the Chief of the Forest Service. VFSYU 1: 67.
5 Ibid.
6 See *Peabody v. US*, 175 U.S. 546, 1899 and *U.S. v. Pena*, 175 U.S. 500, 1899.
7 27 March 1947 Case Study. VFSYU 1: 3A, 21.
8 4 June 1946 letter from Connery and Hood to Carson Supervisor. VFSYU 1: 5.
9 27 March 1947 Case study. VFSYU 1: 3A, 21.
10 17 March 1948 letter from Connery to Carson Supervisor. VFSYU 1: 91.
11 9 March 1953 revised policy statement. VFSYU 2: 186.
12 29 June 1955 letter from Jackson to Carson Supervisor. VFSYU 2: 240.
13 17 February 1953 letter from Regional Forester Lindh to the office of New Mexico Senator Chavez. VFSYU 2: 182.
14 11 June 1955 Memo from District Ranger Starkey. VFSYU 2: 228.
15 14 March 1953 letter from logging workers to the office of Senator Chavez. VFSYU 2: 193.
16 8 July 1955 memo from District Ranger Starkey to Carson Supervisor. VFSYU 2: 237.
17 19 October 1955 report to the Chief of the Forest Service by Assistant Regional Forester Kirkpatrick. VFSYU 3: 280

18 27 May 1957 memo from District Ranger Starkey to Carson Supervisor. VFSYU 4: 412.
19 10 February 1956 letter from the El Rito District to Carson Supervisor. VFSYU 3: 312.
20 10 November 1955 ARO fitness hearing transcript. VFSYU 3: 301.
21 10 November 1955 ARO fitness hearing transcript. VFSYU 3: 301.
22 Ibid.
23 7 November 1966 Plan for Abandonment prepared by District Ranger Miller. VFSYU 4: 447.
24 5 January 1967 letter from Weinstein to Garcia. VFSYU 4: 449.
25 27 January 1968 letter from Weinstein to District Ranger Miller. VFSYU 4: 465.
26 23 March 1967 memo from Regional Forester Hurst to Carson Supervisor Seaman. VFSYU 4: 450.
27 4 February 1971 letter from District Ranger Deiter to Duke City. VFSYU 4: 492.
28 Ibid.
29 4 March 1971 memo from Carson Timber Officer Hutt to Carson Supervisor. VFSYU 4: 495.
30 Ibid.
31 19 February 1971 letter from Deiter to Duke City. VFSYU 4: 499.
32 13 April 1971 memo from Carson Supervisor Hassell to the El Rito Ranger District. VFSYU 4: 504.
33 12 September 1974 letter from Jaramillo to Weinstein. VFSYU 5: 547.
34 18 September 1974 letter from Weinstein to Carson Supervisor. VFSYU 5: 547
35 8 December 1977 memo from Carson Forest staffer Struthers. VFSYU 5: 586.
36 5 May 1986 Carson Forest memo on file at the El Rito Ranger District.

References

Clary, D. (1986) *Timber and the Forest Service*, Lawrence, KS: University Press of Kansas.
Cronon, W. (1991) *Nature's Metropolis: Chicago and the Great West,* New York: W.W. Norton and Company.
Ebright, M. (1994) *Land Grants and Lawsuits*, Albuquerque, NM: University of New Mexico Press.
Ellis, R. (ed.) (1975) *New Mexico Historic Documents,* Albuquerque, NM: University of New Mexico Press.
Gardner, R. (1971) *Grito!: Reies Tijerina and the New Mexico Land Grant War of 1967*, New York: Harper & Row Publishers.
Krahl, L. and Henderson, D. (1998) "Uncertain Steps toward Community Forestry: A Case Study in Northern New Mexico," *Natural Resources Journal* 38: 53–84.
O'Connor, J. (1998) *Natural Causes: Essays in Ecological Marxism*, New York: Guilford Press.
Polanyi, K. ([1944] 1957) *The Great Transformation*, Boston, MA: Beacon Press.
Poling, S. and Kasdan, A. (2001) "Treaty of Guadalupe Hidalgo: Definition and List of Community Land Grants in New Mexico," Exposure Draft, *Report of the United States General Accounting Office*, Washington, DC: US General Accounting Office.
Robbins, W. (1987) "Lumber Production and Community Stability," *Journal of Forest History* 31: 187–96.
Robertson, M. (2000) "No Net Loss: Wetland Restoration and the Incomplete Capitalization of Nature," *Antipode* 32: 463–93.

Rosenbaum, R. and Larson, R. (1987) "Mexican Resistance to the Expropriation of Grant Lands in New Mexico," in C. L. Briggs and J. R. V. Ness (eds.) *Land, Water, and Culture: New Perspectives on Hispanic Land Grants*, Albuquerque, NM: University of New Mexico Press.

Scheper-Hughes, N. (2000) "The Global Traffic in Human Organs," *Current Anthropology* 41: 191–223.

Swyngedouw, E. (1997) "Power, Nature, and the City: The Conquest of Water and the Political Ecology of Urbanization in Guayaquil, Ecuador: 1880–1990," *Environment and Planning A* 29: 311.

Wilmsen, C. (2001) "Sustained Yield Recast: The Politics of Sustainability in Vallecitos, New Mexico," *Society and Natural Resources* 14: 193–207.

20 Neoliberalism and the struggle for land in Brazil

Wendy Wolford

The 1990s was the decade of neoliberalism in Brazil. Public sectors once considered strategic were privatized, social programs were cut back or withdrawn, trade regulations were slashed, and the economy was opened to foreign investment and imports. These policies, euphemistically referred to as international "market integration" (Cardoso 1995, 1999; see also Alimonda 2000), were defended within Brazil and without as necessary to counteract the massive public debt that had accumulated during previous decades of rapid state-led growth (Bresser Pereira 1996; de Onis 2000; Goertzel 1999).

In Brazil, as in many other Latin American countries, neoliberal policies were articulated with the return to democratic rule, after decades of authoritarian dictatorship. Perhaps ironically, the neoliberal emphasis on a leaner state was accompanied by the mobilization of civil society and renewed demands that the central state intervene in economic and political affairs for the purposes of promoting social justice (see especially Dagnino 2002; also Avritzer 2002). One of the most aggressive demands made of neoliberal democracy at this time was the demand for agrarian reform (Medeiros, 1998; Novaes, 1998). Led by the formation of new social movements such as the Movement of Rural Landless Workers (the MST), the rural poor throughout Brazil invoked Article 186 of the Brazilian Constitution to argue for the right to property that was defined as "unproductive," and therefore not fulfilling its responsibility to the broader "social good" (see Fernandes 1999; Gohn 1997; Novaes 1998; Wright and Wolford 2003).

By the end of the 1990s, there was widespread support for agrarian reform in Brazil: middle-class urban Brazilians supported land reform and the MST in astonishingly high numbers (85 percent in one poll), and even rural elites admitted that distribution was necessary because of the historical inequalities in land ownership (see Wolford 2005). What few people agreed upon was how agrarian reform should actually be executed.

The two main perspectives on the issue could be characterized as neoliberal and neo-populist, respectively. The first argued that reform ought to operate through the market (typically if prosaically called Market Led Agrarian Reform or MLAR, see Borras 2003; Deininger 2001), focusing on "willing buyers and willing sellers." This perspective situated claims to land in historical notions of rightful access through hard work, individualism, competitiveness, and playing by the "rules" (of the market); it was generally supported by the Brazilian government, rural elites, and international development agencies such as the World Bank. The second perspective on land reform, closely tied to the MST, argued that land reform ought to operate through the state (typically referred to as state-led agrarian reform or SLAR), where the state would expropriate land, with or without compensation, and distribute it to the poor. Contrary to the neoliberal approach, actors such as the MST situated their claims to land in historical notions of access through hard work, the grace of a socially just God, and collective action – or, "land to those who work it (and need it)" (see also Martins 2000).[1]

In this chapter, I describe the rise of neoliberalism and the struggle for land in Brazil and then analyze the competing perspectives on agrarian reform. I suggest that the Brazilian state has promoted MLAR as being more appropriately articulated with the broader agenda of neoliberal reform. Interviews cited in this chapter were conducted in Brazil by the author in 1998–99.

The 1990s: the neoliberal decade

The increasingly neoliberal economic policies enacted during the 1990s in Brazil marked a dramatic turning point after six decades of protectionism and state "mid-wifery" (Evans 1979). Under Fernando Collor (president from 1990 to 1992), neoliberal reforms including currency stabilization, tariff reduction and active regional market integration came to be seen as the "only game in town" (Nylen 1993). Although Collor's presidency ended early amid corruption scandals and civil mobilization for his impeachment, his early reforms led to the establishment of a regional customs union (Mercosul), creating virtually free trade between Brazil, Argentina, Paraguay and Uruguay in 1995. From 1987 to 1995, trade as a percentage of Brazil's GDP rose from 17 percent to 27 percent and the country's trade-weighted average tariff fell from 51 to 14 (figures cited in Baker 2002).

It was in 1995, however, that Fernando Henrique Cardoso assumed the presidency and neoliberal policies became truly dominant in Brazil (Green 2000). In the 1970s, Cardoso was a respected figure of the Latin American Left. He was well known for having articulated a modified dependency theory, referred to as "associated dependent development" (Cardoso and Faletto 1978), arguing that the timing of Brazil's entrance into the world economy reproduced its dependent condition, even though it allowed for limited sectoral growth. In the 1990s, however, Cardoso campaigned for the presidency on the basis of a neoliberal "paradigm shift" (Power 1998: 51), a project that drew on the strength of his prior currency stabilization plan, the *Plano Real* (or, the Real Plan). Implemented in 1994, the Real Plan was immensely popular because it pegged the Brazilian currency to the dollar and put an immediate end to the hyper-inflation that had plagued the country since the late 1980s (Amman and Baer 2000). Under the plan, controls on bank lending were also put into place to reduce the inflationary practice of state banks lending to the federal government. Price stability and currency appreciation led to increased foreign imports and investment. Foreign direct investment in Brazil increased from less than US$1 billion in 1991 (net inflows) to a high of US$30 billion in 1999.[2] Part of this increase was due to continued regional market integration, primarily through Mercosul and other Latin American countries, but it was also due to the privatization of public utilities and resource sectors such as the steel industry and the Companhia do Vale Rio Doce (CVRD), one of the richest mineral reserves in the world. Between 1995 and 1998, privatization initiatives generated approximately US$60 billion for federal and state governments in Brazil (Sonntag 2002: 88). In promoting these policies, Cardoso argued that "the faith [of the 1960s and 1970s] in all-encompassing and ideological solutions has been lost" (Cardoso 1999: 44) and neoliberalism offered the only viable alternative (Cardoso 1996).

For all of his efforts, Cardoso was hailed by the international financial community as executing the "first stages of a modern capitalist reorganization" (de Onis 2000). The conservative British weekly, *The Economist*, regularly applauded Cardoso's firm commitment to privatization, comparing him favorably to England's Margaret Thatcher. Cardoso was extremely popular throughout his first administration and, after succeeding in having the Constitution modified to allow a second term, was easily re-elected in the first round of the 1998 elections.

Cardoso and the awkward issue of agrarian reform

One of Cardoso's greatest challenges during his presidency was the struggle over land distribution (Pereira 2003; Sorj 1998). When first campaigning for the presidency in 1994, Cardoso promised to address the issue of agrarian reform by settling 280,000 families during his first four-year term. This was more than the number of families settled by all previous federal government administrations combined. Despite these campaign promises, it was evident that economic stabilization and market-oriented reforms – not agrarian reform – were more central to Cardoso's political agenda (Ondetti 2001; Pereira 2003). Cardoso seemed annoyed by the public support for land distribution, calling it a "nineteenth-century demand" (cited in Pereira 2003: 49), and the budget for agrarian reform was widely considered insufficient to settle even 10,000 families. It was during his administration, however, that agrarian reform would become an imperative political issue.

In August 1995, a landless squatter encampment in the municipality of Corumbiara, Rondônia, was attacked by military police and 10 landless squatters were killed, some of them very clearly executed by the police. In response to the public outcry following the incident, Cardoso appointed his personal advisor, Francisco Graziano Neto, to be head of the federal land reform agency, INCRA (the National Institute of Colonization and Agrarian Reform) and promised to increase land expropriations. Less than a year later, in April 1996, a group of 1,200 landless squatters were marching from their encampment to the capital city of the state of Pará along the state highway. They were surrounded by military police who opened fire, killing 19. This time the incident was caught on tape by a local news reporter and the ensuing media coverage caused a national and international scandal. These two incidents were responsible in part for the increased visibility of rural poverty, violence, and the MST (Ondetti 2001).

Cardoso responded publicly and immediately. Twelve days after the massacre, he created a new Extraordinary Ministry of Land Tenure Politics (MEPF Ministério Extraordinário de Política Fundiária) and increased the rate of land expropriations. In 1997, Cardoso settled 80,000 families, almost two times as many as were settled in his first year in office (Cardoso 1999), and from 1994 to 1998, the annual budget for INCRA, the National Institute for Colonization and Agrarian Reform, was more than quintupled (Seligmann 1998).[3]

Market-led agrarian reform

At the same time as Cardoso expanded the state-led program of agrarian reform, however, he also implemented a new market-based approach to agrarian reform in keeping with his overall neoliberal policy reforms. From 1998 to 2003, President Fernando Henrique Cardoso championed a Market-Led Agrarian Reform (MLAR) that would distribute land from willing sellers to willing buyers with as little government intervention as possible (Borras 2003). This "new model of land policy [was to be] integrated into the market and independent of the government at each stage of the process" (Cardoso 1995). Reliance on the state had come to be seen as mixing economics with politics: even though most studies on land tenure patterns in Brazil argue that inequality in land ownership has been supported (if not created) by the state (Holston 1991), large landowners and government officials in Brazil argued that pursuing re-distribution through the government was "forcing the issue" and an unacceptable "way out." The large farmers considered the MST's reliance on state intervention to be evidence itself that the rural poor were attempting to evade the difficult work of making an honest living. As one farmer said, "If they [the MST members] were farmers, really, they would [already] be producing. These people around here who are holding meetings in the peripheries [of the city] have never held a hoe, they're only there to make trouble. They are thieves, they are thugs."

International financial and development agencies such as the World Bank supported the implementation of MLARs around the Third World in the

1990s, arguing that they were both more efficient and less expensive than state-led reforms (Borras Jr. 2003; De Janvry *et al.* 2001). With an initial loan of US$90 million, the World Bank established a pilot project called A Cedula da Terra (The Land Title) in 1997 in five northeastern states (Ceará, Pernambuco, Maranhão, Bahia and Minas Gerais). The Cedula da Terra was the forerunner of a broader market-led agrarian reform (MLAR) project, called O Banco da Terra (The Land Bank). In 1998, the Land Bank became an official program and was organized in collaboration with federal and state government funds. Both the Cedula da Terra and the Banco da Terra targeted people who had experience in subsistence agriculture and whose annual income did not exceed US$15,000 or who did not already own a property that was larger than a "family farm" as defined by local conditions. These "rural producers" were required to form associations with other interested buyers, and all negotiations over property were voluntary. In this way, land re-distribution would follow the market logic of supply and demand: people who needed land would find land that needed people. This logic was reiterated by large farmers who supported the MLAR:

> Agrarian reform is necessary. But this [the MST's way] is not the way to do it ... If I don't have land and I want land to work on, then I have to put my head down and go where there is land available – in Mato Grosso. I have to go where there is public land ... You have to go where the work is ... [You can't say] "I want land here because I am Brazilian, because I am a poor little guy (*coitadinho*), because I am a worker, because ... " It isn't like this! We have to go where the work and land exist and not demand that it be given to us here.

According to the rules of the MLAR, the participants were provided with loans of up to US$40,000 to help them purchase land. As "rural producers," they would receive state-subsidized assistance in establishing local infrastructure, and then they would have 20 years to pay back the loan, with a grace period of three years before interest rates of between 4 and 6 percent applied.

The MST: "Land for those who work it"

> God didn't sell the land to anyone, he left it for us. In the time of my parents, land was not sold, you just went there.
>   (MST member and landless farmer, 1998, interview with the author)

In 1984, approximately 400 rural squatters came together to form the MST. The MST's success in building an organized social movement was a product of several factors, including the changing political environment, the increased landlessness due to the ongoing modernization of agriculture, mobilization assistance provided by the Catholic and Lutheran Churches, and the appeal to a group of people whose cultural practices of production generated a desire and need for continued access to land (Wolford 2003a). The MST's main tactic was the direct-action land occupation: movement recruits and activists selected and then occupied a large property – usually one defined legally as unproductive. MST members then squatted on the land to force the government to recognize their claim to its productive use. If the occupation was successful, the government expropriated the property and divided it among the landless poor. As of June 2004, the MST had helped to establish well over 1,000 land reform settlements and claimed to represent almost two million members throughout the country.

The MST's claim to "land for those who work it" was fundamentally different from (and opposed to) the neoliberal logic of the MLAR. Instead of beginning with an assumption of the equality of markets (or the equalizing ability of markets), the MST's neo-populist perspective began with the assumption of inequality, the belief that landlessness was not "natural" nor an indication of incompetence and laziness, but rather was an indication of the skewed distribution of state resources (agricultural subsidies and credit primarily) and the extent to which land had historically been made available primarily for those with the financial resources to produce export commodities. MST members argued that land ought to be distributed through the state rather than through the market because of those original inequalities and because the two institutions would promote two very different meanings of which "unproductive" land could be distributed: within the language of the state, included in the Federal Constitution, "unproductive" meant that the land was not producing sufficiently to guarantee its "social function," meaning that all arable land had an obligation to produce certain quantities of food or cattle as long as there were people going hungry or in need of land; within the language of the market, however, "unproductive" meant land not being demanded by its own owner. The MST argued that in order for the landless to receive good-quality land, they would require state intervention. The movement also argued that the demand for land should not be forced to follow the supply, rather, the state should supply land where it was demanded, essentially in the home communities where the landless were familiar with local (agri)cultural conditions.

The MST's neo-populist perspective emphasized the centrality of land, community, and the local, where all were believed to be key to both production and social reproduction and where farmers who produced for their families were the proper stewards of the material environment. In this framework, land is not just a material asset, it is the key to a meaningful livelihood and to effective citizenship in Brazil; or, as one MST activist said: "Land is life." Access to land signified access to stability, security, a "place of one's own." The land was more than employment or food, it was home, and it was history: "Land means a lot – that's where your life is. I was born on the land ... [and] all I know how to do is work on the land. On the land you don't go hungry."

Within this perspective, land was also the means to building community. The notion of community was based on and in the traditions of farming communities where each person was tied to the land and to each other through relatively non-commercial bonds of solidarity. Small farmer communities were shaped by production relations: working and living on the land created common interests (the weekend soccer game was mentioned regularly) as well as the need for occasional cooperation. Short-term work-parties (*mutirões*) were held in farming communities in southern Brazil when urgent or unwieldy tasks arose, such as the need to build a school-house or repair a neighbor's barn (Wolford 2004).

Embedded in the notion of access to land and community was also the right to local sovereignty, particularly food sovereignty, an argument that MST leaders members made at different scales, from the local to the national and international. Food sovereignty implied local control over food from the point of production to consumption, and the MST opposed this economic model to the logic of Brazil's agro-industrial corporations that had achieved international success exporting fruits, meats, and grains around the world. In response to the argument that MST members could not effectively compete with large farmers, MST members argued that they were the more efficient producers if one considered the end to be feeding hungry people, providing healthy food, and practicing sustainable production methods rather than earning foreign currency.

The fight for local control over food led the MST to wage an aggressive campaign against what movement leaders saw as the privatization of the food supply, particularly in the genetically modified seeds being pushed (equally aggressively) by large multi-national corporations such as Monsanto. Movement members argued, along with many other farmers' movements around the world, that by creating and patenting new seed varieties that are incapable of reproducing and require specific inputs, corporations threaten the control individual farmers have over production decisions. All of this – the emphasis on community, land, and food sovereignty – stood in discursive opposition to the society and economy associated with the neo-liberal state and market.

The MLAR and SLAR, in theory and in practice

In principle, the MLAR and the SLAR were promoted by the Brazilian state as distinct and complementary. The MLAR would target small and medium-sized farmers who were willing to sell their land and who were able to negotiate an acceptable price with the landless, while the SLAR would target large-scale latifundia. In practice, however, when the Land Bank was established in 1998, the government began to withdraw resources from the state-led agrarian reform process. In 1997, when the Cedula da Terra began, INCRA's annual budget was R$2.6 billion and by 2001, it was roughly half that amount. From 2001 to 2003, the official budget for state-led agrarian reform was cut by a further 39 percent. The government's overall agenda was characterized by its reliance on decentralization, processual privatization, and marketization. In March 2000, Raul Jungmann, head of INCRA from 1997 to 2002, presented the outline for the *Novo Mundo Rural* (the New Rural World) that included (a) decentralization of agrarian reform with emphasis on state and municipal level collaborations; (b) the elimination of special credit geared towards settlers; (c) the incorporation of land reform settlers into the same government program as small farmers; and (d) expediting the process of "liberating" land reform settlers from their dependence on the state.

The government's rhetoric supported the idea that a market-led reform would be more efficient and targeted towards more appropriate beneficiaries than a state-led reform. In a 1999 document called *The Land Bank*, the government argued that the MLAR would be a success because it forced the landless farmer to take personal responsibility for his future:

> [the landless farmers] buy the land in cash and have 20 years to pay the loan. It is good land that they chose by themselves. With the help of the [agricultural extension agents], they negotiate the price with the [landowners] until they get the best offer [possible].
>
> (MDA 1999: 7)

As a "poor but rational" consumer and producer, the landless person who works through the Land Bank is "not a passive agent, a non-participant in [an otherwise] administrative process" (MDA 1999: 26), they are "rural producers," not land reform beneficiaries, so they do not rely on the state for assistance. The government document proudly cites one land recipient as saying: "Nobody here is going to ask [for] anything from City Hall. We are working and producing, not begging anyone" (MDA 1999: 38).

Behind this putatively rational and value-free logic of MLAR lay a very political attempt to discredit the logic of an alternative approach, namely that of the SLAR. In describing the advantages of purchasing land through the MLAR, the government directly and indirectly compared Land Bank recipients with MST settlers, and it was clear who won and who lost.[4] At the end of the document on the Land Bank, an example is given of a land sale that was almost overturned because of the MST:

> The president of the association, Espedito Augusto da Luz, [explains how] the negotiation for the purchase of the farm got so far behind. "On the eve of the [deal closing] with the former owner, the farm was invaded by the Landless Movement (the MST). They [the MST] left quickly but they [set up an occupation camp] beside the entrance [to the farm], and the negotiations crawled."
>
> (MDA 1999: 41)

Another Land Bank participant affirmed, "[the Land Bank] is better than the normal projects [state-led agrarian reform] because it involves people who always lived off the land. In the invasions, we see a lot of people who don't have a history of connection with the land" (ibid.).

The neoliberal logic of the MLAR outlined and was supported by a world-view that paralleled that of the agrarian elite (Wolford 2005). It situated the traditional rights to land in hard work, personal responsibility, and reliance on the market rather than on "politics," where engaging in politics was seen as a lowly form of begging. It defended rights guaranteed by the market (property rights, consumer rights), rather than rights guaranteed by the state or civil society (human rights, social rights, or the "right to have rights"). Ultimately, well-being was determined by the "laws" of supply and demand, and the "forces of competition" rather than by subjective proscriptions for social justice or equality. This moral reasoning – progress through hard work – provided an important mechanism for interpreting a particular person's landlessness or poverty as a failing of the *individual*, even though in the abstract landlessness and poverty were recognized as difficult societal problems.

Acknowledgments

A longer version of this chapter was Wolford, W. (2005) "Agrarian Moral Economies and Neo-liberalism in Brazil: Competing World-Views and the State in the Struggle for Land," *Environment and Planning A* 37: 241–61.

Notes

1 At a discursive level, the MST acts with a unified, coherent voice, representing the interests and wishes of its considerable membership. In truth, of course, the "movement" is not always unified, rarely coherent, and represents only a partial set of its members' beliefs. In other work, I have examined the inner workings of the movement more closely, but for the purposes of this chapter, the analysis will focus on the MST at the organizational level.
2 Figures from the Central Bank, cited on the Embassy of Brazil (London) web page at: http://www.brazil.org.uk/page.php?cid=1170 (accessed November 25, 2003).
3 In 1994, INCRA's annual budget was R$390 million, and in 1998, the agency's annual budget was R$2,243 million (Seligmann 1998).
4 In the six years since the establishment of the first Land Bank projects, the academic evaluations have been mixed. For positive evaluations, see Deininger (2001). For negative evaluations, see Navarro (1999) and Borras Jr. (2003).

References

Alimonda, H. (2000) "Brazilian Society and Regional Integration," *Latin American Perspectives* 27(6): 27–44.
Almeida, L.F. de and Sanchez, F. R. (2000) "The Landless Workers' Movement and Social Struggles against Neoliberalism," *Latin American Perspectives* 27(5): 11–32.

Amman, E. and Baer, W. (2000) "The Illusion of Stability: The Brazilian Economy under Cardoso," *World Development* 28(10): 1805–19.

Avritzer, L. (2002) *Democracy and the Public Space in Latin America*, Princeton, NJ: Princeton University Press.

Baker, A. (2002) "Reformas liberalizantes e aprovação presidencial: a politização dos debates da política econômica no Brasil [Free-Market Reform and Presidential Approval: The Politicization of Economic Policy Debates in Brazil]," *Dados* 45(1): 77–98.

Barham, B. and Carter, M. (1996) "Level Playing Fields and Laissez Faire: Post-liberal Development Strategy in Inegalitarian Agrarian Economies," *World Development* 24(7): 1133–49.

Borras Jr., S. M. (2003) "Questioning Market-Led Agrarian Reform: Experiences from Brazil, Colombia and South Africa," *Journal of Agrarian Change* 3(3): 367–94.

Bresser Pereira, L. C. (1996) *Economic Crisis and State Reform in Brazil: Toward a New Interpretation of Latin America*, Boulder, CO: Lynne Reiner Publishers.

Cardoso, F. H. (1995) "Brazil and Current Challenges," a speech given at the National Press Club, Washington, DC April 21.

—— (1996) "Globalização," speech given at the Universidad Autonoma de Mexico and reprinted in the *Folha de São Paulo*, February 2, 1996.

—— (1997) *Reforma Agrária: Compromisso de Todos*, Brasília, DF: Presidência da República Secretaria de Comunicação Social.

—— (1999) "From Dependencia to Shared Prosperity," *New Perspectives Quarterly* 95(12/1): 42–4.

Cardoso, F. H. and Faletto, E. (1978) *Dependency and Development in Latin America*, Berkeley, CA: University of California Press.

Da Veiga, J. E. (1990) *A reforma que virou suco: uma introdução ao dilema agrário no Brasil*, Petrópolis: Vozes.

Dagnino, E. (1998) "Culture, Citizenship, and Democracy: Changing Discourses and Practices of the Latin American Left," in S. E. Alvarez, E. Dagnino and A. Escobar (eds.) *Cultures of Politics/Politics of Culture: Re-Visioning Latin American Social Movements*, Boulder, CO: Westview Press.

—— (2002) "Sociedade Civil e Espaços Publicos no Brasil," in E. Dagnino (ed.) *Sociedade Civil e Espaços Publicos no Brasil*, São Paulo: Editora Paz e Terra S/A.

De Janvry, A., Sadoulet, E. and Wolford, W. (2001) "The Changing Role of the State in Latin American Land Reforms," in A. de Janvry, G. Gordillo, J-P. Platteau, and E. Sadoulet (eds.) *Access to Land, Rural Poverty, and Public Action*, Oxford: Oxford University Press.

De Onis, J. (2000) "Brazil New Capitalism," *Foreign Affairs* (May/June).

Deininger, K. (2001) "Negotiated Land Reform as One Way of Land Access: Experiences from Colombia, Brazil and South Africa," in A. de Janvry, G. Gordillo, J-P. Platteau, and E. Sadoulet (eds.) *Access to Land, Rural Poverty, and Public Action*, Oxford: Oxford University Press.

Evans, P. (1979) *Dependent Development: The Alliance of Multinational, State, and Local Capital in Brazil*, Princeton, NJ: Princeton University Press.

Fernandes, B. M. (1999) *MST, Movimento Dos Trabalhadores Rurais Sem-Terra: Formação e Territorialização*, São Paulo: Editora Hucitec.

Goertzel, T. G. (1999) *Fernando Henrique Cardoso: Reinventing Democracy in Brazil*, Boulder, CO: Lynne Reiner Publishers.

Gohn, M. d. G. (1997) *Os Sem Terra, ONGs e Cidadania*, São Paulo: Cortez.

Graziano da Silva, J. F. (1982). *A Modernização Dolorosa: Estrutura Agrária, Fronteira Agrícola e Trabalhadores Rurais no Brasil*, Rio de Janeiro: Zahar Editores.

Green, J. N. (2000) "Introduction," special issue of *Latin American Perspectives*, Brazil in the Aftershock of Neoliberalism, 27(6): 5–8.

Grindle, M. (1986) *State and Countryside: Development Policy and Agrarian Politics in Latin America*, Baltimore, MD: Johns Hopkins University Press.

Hammond, J. (1999) "Law and Disorder: The Brazilian Landless Farmworkers' Movement," *Bulletin of Latin American Research* 18(4): 469–89.

Holston, J. (1991) "The Misrule of Law: Land and Usurpation in Brazil," *Comparative Studies in Society and History* 33(4): 695–725.

McCarthy, J. P. (1998) "The Good, the Bad, and the Ugly: Environmentalism, Wise Use, and the Nature of Accumulation in the Rural West," in B. Braun and N. Castree (eds.) *Remaking Reality: Nature at the Millennium*, London: Routledge.

Martins, M. D. (2000) "The MST Challenge to Neoliberalism," *Latin American Perspectives* 27(5): 33–45.

Medeiros, L. (1998) "Os trabalhadores rurais na política: o papel da imprensa partidária na çãio de uma linguagem de classe," in L. F. C. Costa and R. Santos (eds) *Política Reforma Agrária*, Rio de Janeiro: Mauad.

Ministério do Desenvolvimento Agrário (MDA) (1999) *The Land Bank*. English translation of the original document: *O Banco da Terra*. Brasília: Instituto Nacional de Colonização e Reforma Agrária.

Müller, G. (1985) *A dinâmica da agricultura paulista*, São Paulo: Fundação Sistema Estadual de Análise de Dados.

Navarro, Z. (1999) "The 'Cedula da Terra' Guiding Project: Comments on the Social and Political-Institutional Conditions of its Recent Development," Washington, DC: The World Bank.

Novaes, R. (1998) "A trajetória de uma bandeira de luta," in L. F. C. Costa and R. Santos (eds) *Política e Reforma Agrária*, Rio de Janeiro: Manad.

Nylen, W. R. (1993) "Selling Neoliberalism: Brazil's Instituto Liberal," *Journal of Latin American Studies* 25(2): 301–11.

Oliveira, P. C. and Del Campo, C. P. (1985) *A propriedade privada e a livre iniciativa no tufão agro-reformista*, São Paulo, SP: Editora Vera Cruz Ltda.

Ondetti, G. (2001) "When Repression Backfires: The Rise of the Brazilian Landless Movement in the Mid-1990s," paper presented at the Latin American Studies Association meetings in Washington, DC.

Pereira, A. W. (2003) "Brazil's Agrarian Reform: Democratic Innovation or Oligarchic Exclusion Redux?" *Latin American Politics and Society* 45(2): 41–65.

Power, T. (1998) "Brazilian Politicians and Neoliberalism: Mapping Support for the Cardoso Reforms, 1995–97," *Journal of Interamerican Studies and World Affairs* 40(4): 51–72.

Seligmann, R. (1998) *PROCERA: Programa Especial de Credito para Reforma Agrária*, Brasília, DF: Instituto Nacional de Colonização e Reforma Agrária.

Sonntag, H. R. (2002) "Toward the Core of the World System? Social Change in Brazil," *The Brown Journal of World Affairs* 8(1): 83–96.

Sorj, B. (1998) "A reforma agrária em tempos de democracia e globalização," *Novos Estudos* 50: 23–40.

The Economist (1998) "Brazil's Steady Nerve," October 10, pp. 6–10.

Wolford, W. (2003a) "Families, Fields, and Fighting for Land: The Spatial Dynamics of Contention in Rural Brazil," *Mobilization: An International Journal* 8(2): 157–73.

—— (2003b) "Producing Community: The MST and Land Reform Settlements in Brazil," *Journal of Agrarian Change* 3(4): 500–20.

—— (2004) "Of Land and Labor: Agrarian Reform on the Sugarcane Plantations in Northeast Brazil," *Latin American Perspectives* 31(2): 147–70.

—— (2005) "Agrarian Moral Economies and Neo-liberalism in Brazil: Competing Discourses and the State in the Struggle for Land," *Environment and Planning A* 37: 241–61.

Wright, A. and Wolford, W. (2003) *To Inherit the Earth: The Landless Movement and the Struggle for a New Brazil*, Oakland, CA: Food First Publications.

21 Enclosure and economic identity in New England fisheries

Kevin St. Martin

Introduction

I am drawn to thinking about the commons because of its ability to define a place as outside of capitalism; I am enticed by stories of societies and environments and their myriad productive combinations before capitalism; and I am inspired to imagine alternative ways of being that real people have lived and are living on the commons. I am, however, frustrated by representations of the commons as always subject to an inevitable displacement by a dominant and invasive capitalism (Gibson-Graham and Ruccio 2001). It would seem that all of our stories of the commons revolve around a capitalist imaginary: capitalism's origin in the enclosure of the commons, capitalism's commodification of natural resources, capitalism's expansion and its penetration of common property regimes globally, and capitalism's most recent push to privatize remaining common property resources via neoliberal policies at a variety of scales (Community Economies Collective 2001). A commons future is difficult to imagine.

How this problematic of representing the commons is enacted in a contemporary common property regime, marine fisheries of New England, is examined in this chapter. Here, the problematic is clearly evident in the narrow range of solutions available to address environmental and industrial crises. Fisheries in New England (and around the world) are being gradually but inevitably privatized (commodified, marketized, etc.) in an effort to place them within the domain of capitalism where private rights to resources will ensure an attitude of stewardship amongst capitalists and, as the dominant ideology would have it, a long-term environmental sustainability (Eckert 1979; Garcia *et al.* 1999; Hannesson 2004). While many doubt the promise of privatization will be realized (in either social or environmental terms), any evidence that fisheries might be alternatively managed by, for example, communities or within community-based economies is dismissed via its relegation to the status of a vestige. Processes that might suggest other futures are possible are read as remnants of a pre-capitalist past where fishermen *were* embedded in communities, territories, and a socially produced economy (cf. Callari 2004).

While represented as archaic, distant, or subordinate, the commons nevertheless remains a powerful metaphor for alternative forms of human and environmental organization (McCarthy 2005). As a spatial metaphor for economic difference, I want to suggest that we reexamine the potential of the commons as a contemporary location of a diverse economy harboring multiple economic possibilities (cf. Gibson-Graham 2006; Gibson-Graham *et al.* 2000; Leyshon *et al.* 2003; Pavlovskaya 2004). In so doing we should not ignore processes of enclosure and their clear ability to transform economies, societies, and environments, but nor should we concede the entirety and the rhetorical power of the commons to a narrative of capitalist enclosure and capitalist development.

Below, I will suggest that the on-going enclosure of fisheries in New England is the result of "capitalocentric" representations of the commons that make alternative solutions to commons problems difficult to imagine (Gibson-Graham 1996). While the commons of fisheries in New England is a degraded (even tragic) environment subject to industrial overcapitalization and increased fishing effort over time due to the investment strategies of individual boat owners, it is also a commons filled with and constituted by community relations, community-economic processes, territorializations, and local understandings and representations of resources (St. Martin 2001).

Representing a fisheries commons

The following is based upon results from a set of semi-structured interviews with fishermen[1] in Massachusetts as well as several years of participant observation of fisheries science and management in New England (e.g. attending fisheries management council meetings, participating on scientific and management committees, etc.). The research is, however, presented via the story of Bob, a key informant who was interviewed four times while he was working out of Plymouth, Massachusetts, at the time of the interviews (1998/99).

The interviews were designed to document the processes of community and territory in which fishermen might participate. The goal was to locate and make evident processes that were thought not to exist in the industrialized fisheries of New England, to uncover community-based identities and spatial processes reminiscent of a pre-capitalist commons and hence an opening for community-based forms of resource use and economy. These discoveries, it was hoped, would replace/correct the assumed individual identity and open access nature of fisheries that are more commonly ascribed to this location and that point to a necessary enclosure and capitalist solution to the ongoing environmental crisis (St. Martin 2001). Examining, in detail, the practices of a single fisherman, however, suggests that identity and spatial understandings of the commons do not fit easily into a capitalist/pre-capitalist binary with its respective proscriptions for resource management.

Bob's breadth of experience in the fisheries of Southern New England made him an ideal candidate for an enquiry into questions of fishermen's identity as well as their spatial practices. Bob had extensive experience on both offshore and inshore boats; he held a number of positions throughout his career (deckhand, engineer, mate, and captain); he now owned and operated an inshore boat crewed by himself and a single deckhand; and he traveled seasonally to different fishing grounds and interacted with several communities of fishermen. He was moderately successful and well respected by other fishermen in Plymouth and other nearby ports.

Bob owns a 65-foot trawler and focuses on cod and flounder when fishing from Plymouth. Unlike many of the other boats that tie up in Plymouth, Bob's "homeport" is not Plymouth. That is, he fishes from Plymouth only seasonally and has done so for only the last eight years (at the time of interview). At other times of the year he pursues fluke and squid south of Cape Cod in Vineyard Sound (as do some other Plymouth fishermen) closer to his homeport of New Bedford, Massachusetts. Bob's insights into the fishing community of Plymouth are possible largely because of his initial status as an "outsider." Indeed, the question of community as well as the complex nature of Bob's identity as a fisherman is revealed by his interactions with the other fishermen of Plymouth (see below).

New England fisheries as enclosure

What follows is a standard and somewhat leftist story of the regime in fisheries in New England. While very brief, it captures the tenor of current fisheries management and its impacts upon fishermen such as Bob. While this story serves to provide yet another example of the negative impacts of neoliberal resource management polices and to produce a general indignation relative to capitalism, I am interested here to point to what might be the limitations of remaining within such a narrative. That is, what economic potentials do we forfeit by seeing the fisheries commons of New England as always a location where community and commons-based economies are retreating and capitalism is advancing?

The fishermen of Plymouth, like those throughout New England, have over the last decade seen their access to fish curtailed via a variety of scientifically informed management mechanisms designed to reduce fishing effort by species for the management region as a whole. Moratoria on licenses for particular species, limitations on the numbers of days-at-sea per year (DAS), a variety of gear restrictions, ever-smaller landing limits, and seasonal or spatial closures of fishing grounds are all part of the regulatory regime these fishermen must now navigate in order to survive. The fisheries commons of New England, due to pressures from environmental organizations to comply with federal regulations designed to stop overfishing, has been rapidly transformed from an "open access" resource to one where access to resources is highly regulated and limited at the scale of the management region.

In the case of New England, the thrust of the most recent round of regulation has been to control/reduce fishing effort through restrictions that suggest an incremental privatization as seen in "ownership" of permissible days-at-sea or fishing licenses. Indeed, the sense that today's broad mix of regulations is temporary and that full privatization, via Individual Transferable Quotas (ITQs), is the ultimate solution is palpable across a variety of documents, management meetings, and in interviews with a range of people involved in New England fisheries. Although itself having different forms, ITQs have been implemented in fisheries around the world but most notably in New Zealand, Iceland, Nova Scotia, and select fisheries of the United States (Apostle *et al.* 2002; McCay and Brandt 2001). In these cases, and to varying degrees, access to fish in the form of catch quotas becomes itself a commodity that can be bought and sold on open markets, consolidated by individuals or corporations, and employed anywhere within the management region.

This movement in fisheries can be easily read as a classic enclosure of the commons with implications not unlike the enclosures of agricultural commons in Europe and elsewhere. The negative impacts of enclosure in fisheries, specifically the institutionalization of ITQs, have been predicted and documented by social scientists (Apostle *et al.* 2002; Davis 1996; Palsson and Helgason 1995). In these stories the positive effects of ITQs to limit and stabilize fishing effort (and hence the resource) are counterpoised with the host of social and economic disruptions faced by fishing communities. While ITQs may indeed be a benefit for those who hold the right to access fish, they necessarily remove that right from other fishermen and citizens generally. Such systems are plagued by the threat of consolidation and ownership of fishing rights by individuals or corporations no longer embedded within fishing communities. They can lead to a spatial consolidation of the fishing industry such that smaller ports are abandoned as the industry is rationalized and centralized. As with Marx's recollection of English and Scottish enclosures (Marx 1976), abandoned homes, churches, processing plants, and docks can be found in those ports along the New England coast that are too far from more populated areas for gentrification.

ITQs not only have the potential to consolidate the right to fish, they also suggest a transformation of relations amongst fishermen who work together on individual boats. Currently, all crew onboard most fishing boats work not for wages but an equal share of the catch. This is known as the "lay system" in fishing and is currently widespread throughout New England (Doeringer *et al.* 1986). All crew are legally "co-venturers," which positions them as "fishermen" along with those fishermen who own boats. Once access to fish is given to boat owners rather than, for example, some larger definition of "fisherman," the potential for a deepening division between owners and non-owners seems imminent. The dispossession of the right to fish from non-boat-owning fishermen (co-venturers) represents a "quiet confiscation" (Marx 1976) of the fisheries commons and a potential proletarianization of the majority of fishermen in New England.

In addition to legislative redefinitions of property rights, Marx (1976) makes clear that technological innovations, trade and market mechanisms, the availability of capital for investment, greed, violence, and the performance of a discourse and logic of enclosure also constitute enclosure. A closer examination of this last process is warranted because it is at this level that public policy is legitimated, assumptions about fishermen's identities and behaviors are formulated, and capitalism is produced as the natural and inevitable future for the commons.

Enclosure and capitalist identities

The discourse that legitimates the enclosure of fisheries commons does so by conceiving the commons as essentially the same as other sites of capitalism except for the curious institution of common property. This is perhaps most clearly the case in Hardin's often-referenced article on the "tragedy of the commons" (1968). While the economic subject, space, and dynamic assumed by Hardin are identical to other neoclassical theories of capitalism, they become clearly capitalocentric insofar as the commons is part of a linear trajectory of society, embedded within a story of modernization, technological advance, and population growth that necessitates enclosure as essential to economic development. Like other stages of development theories, Hardin's story contributes to a before and after binary that revolves around a modernist development practice where enclosure becomes the hallmark of a modern, capitalist, economy (Callari 2004). Defined this way, Hardin's story and the story of enclosure in New England fisheries told above share a common belief in the direction of economic transformation (i.e. towards capitalism) despite their divergent politics around this transformation.

In addition to the necessity of enclosure for environmental sustainability and economic development, Hardin also makes explicit the necessity of enclosure relative to a stable and centered capitalist identity. That is, a space where resources are available to all (the open access commons) combined with the desires of the modern/capitalist economic subject produces not only environmental degradation but also a psychological crisis. Torn between good conscience/restraint, which would benefit all, and their desire to abuse the commons for their own individual benefit, individuals are caught in a "double bind" that produces "pathogenic effects" such as guilt and anxiety. The double bind is also

> an important causative factor in the genesis of schizophrenia. The double bind may not always be so damaging, but it always endangers the mental health of anyone to whom it is applied. "A bad conscience," said Nietzsche, "is a kind of illness."
>
> (Hardin 1968: 1246)

The modern and economically rational individual within a commons regime finds their mental health threatened as well as their ability to appropriate resource rent. On the commons, the subject (a natural and immutable economic man) cannot be a whole, centered, and modern economic being. The resolution to this problematic is to admit our denial of the simple truth of the commons (Hardin 1977) and enclose the commons. The goal is to erase contradiction, thereby producing the conditions of a sound environmental, economic, and psychological well-being. Subjectivity, space, and identity are fixed by a technical and discursive enclosure such that they merge into a single dynamic that is clearly recognizable as capitalism.

As the subject of empirical study for several decades, the commons has been discovered to be much more complex than originally theorized by Hardin and other neoclassical common property theorists (Robbins 2004). That is, many institutional studies of contemporary commons have corrected and qualified Hardin's thesis and have produced a rich literature on the variety of commons solutions, ways in which the commons can continue despite the forces of industrialization, modernization, population growth, etc. (Ostrom *et al.* 1994; Schlager and Ostrom 1992). These studies, however, continue to reference the same basic economic ontology as Hardin; the solutions to tragedy are represented as technical solutions that build upon the rational economic choices of individuals to appropriate resources and produce individual wealth (see also Mansfield 2004). Tragedy may be averted, but the commons remains a site of negotiation between essentially individual utility-maximizing subjects who seek to appropriate quantities of resources; the commons remains within the domain of capitalism, an essentially capitalist economic space.

Finally, in much anthropological and political ecology literature the commons is represented not as a capitalist space but, either explicitly or implicitly, as a space of economic difference. On these commons, there is an escape from a narrow capitalist economic identity; there, processes of kin, community, culture, territory, etc. provide alternative bases for economic identities and practices. These representations remain, however, within a capitalocentric discourse insofar as they are stories from locations bounded by capitalism. That is, originating in the "Third World" or other locations that are distinctly peripheral to capitalism (e.g. rural areas and first nation territories), they are relegated to a binary position (in this case, the subordinate, weak, and often literally distant position) relative to the presence of capitalism. Importantly, these representations of the commons offer an imaginary of economic difference but, insofar as capitalism retains its hegemony, they remain unimaginable as solutions to the problems of proximate commons (St. Martin 2005).

Limitations of the binary commons

Returning to the example of Bob, how can he be represented given the binary capitalist/pre-capitalist as it relates to the commons? The standard neoclassical discourse of fisheries positions Bob as an individual economic agent who moves from one utility-maximizing opportunity to another, as seen in Bob's seasonal shifting from one fishing ground and/or species to another. In this movement Bob is unconstrained and is, indeed, legally free to fish anywhere. In addition, he is unconstrained by cultural, territorial, or community based relations or restrictions. That some of his New England neighbors might demonstrate other behaviors based on such constraints (Acheson 1988) is the degree to which they are remnants of pre-modern forms of fishing. Bob's level of capital investment, advanced onboard technologies, and fishing capacity suggest a level of modernization and industrialization that places him beyond any remaining processes of culture, territory, or community.

For similar reasons, it becomes difficult to see Bob as a member of a traditional fishing community. Bob lives in a relatively small coastal town that has a small population of resident fishermen with whom he does not associate. His "homeport" is not where he lives; it is in New Bedford, an historic center for the fishing industry in this region. Also, Bob's seasonal movements place him in different ports at different times of the year and Bob lands his catch in a variety of locations depending upon price. To which "traditional" community does Bob belong? Which of his several fishing locations is the "territory" of his "community"? When asked, Bob could not insert himself into a particular bounded community or territory, nor could he claim membership in any fishermen-related community group, fishermen's union, or fishermen's association.

The capitalocentric story of the commons successfully produces a desire for non-capitalism and a place for its enactment but, at the same time, relegates those desires and locations always to the past. As a result, any fisherman can only be represented as either the archaic subject of a fading pre-capitalism or as the emerging capitalist on the now enclosed commons. Given Bob's testimony and fishing practice, he is most easily situated as having abandoned a "traditional" and community-centered way of life, and is advancing toward capitalism.

Re-presenting the commons

> [T]hey ... redefine space as something that cannot be definitively dedicated to particular activities or exhaustively structured by a single form or "identity," ... [t]his space is open, full of overlaps and inconsistencies, a place of aleatory relations and redefinitions, never fully colonized by the pretensions of a singular identity.[2]

The search for alternatives to neoliberal policy in New England fisheries cannot proceed by fixing Bob's identity as a pre-capitalist subject on the commons (and all that that implies). This untenable position offers little in terms of a viable and progressive politics for Bob and fishermen like him.

An alternative strategy is to disrupt the singularity (and capitalocentrism) of dominant (and subordinate) economic identities and commons spaces, to open up the space of the commons to other economic possibilities. Detailed information concerning the fishing practices of Bob is presented below. It suggests an alternative reading of Bob's identity and behavior.

The ambiguous commons or how to fish successfully

Bob, despite great pressure to produce him as such, is not reducible to the mobile, disembedded, and independent fisherman. Indeed, if we look closely at how fishermen enact mobility, disembeddedness, and independence, we can reveal within these processes their contradictory nature. Bob's ability to fish (or not) in a variety of places and his ambiguous relationship to fishing "communities" are, in part, determined by Bob's sharing of cartographic/ environmental knowledge with other fishermen. That is, we can see in the process of knowledge exchange how Bob's identity as a community member and the degree to which he is mobile and independent can be represented in multiple (and contradictory) ways.

The problem of "search," overlooked by the dominant discourse of fisheries but vital to fishermen's success, is embedded in fishermen's practices of sharing knowledge, in particular, detailed knowledge of the geography of the ocean bottom (Wilson 1990; Palsson 1994). Rocky bottoms provide habitat for groundfish, which are increasingly scarce elsewhere. Rocky bottoms also damage nets, particularly those towed by small and medium-sized boats that, unlike larger boats, do not have crews to mend damaged nets. Success in the inshore areas frequented by Bob and fishermen of Plymouth is largely a function of the detailed knowledge of where a net can be towed near rocky areas without being damaged. Other, less productive, inshore areas, particularly those with smooth or sandy bottoms, are known by most fishermen and no detailed knowledge of the bottom is needed.

In Plymouth, knowledge of the rocky areas is traded amongst fishermen in the form of "papers" produced on plotter machines on board each boat. When the net is cast and towed, the plotter is turned on. It traces the path of the boat until the tow is complete and the net reeled in. The path appears on the paper as a simple line within the Loran coordinate system grid. Successful paths are repeatedly used and repeatedly mapped on a single paper. "Papers" containing good "tows" are often sheets of loose-leaf paper that are easily photocopied and traded amongst fishermen. The result is a community of fishermen who share information about individual or multiple tows in specific locations – tows that determine the success of inshore fishermen.

Papers are traded depending upon relationships between fishermen, the expectations of reciprocity, and the value of the paper being traded. For example, general tows that are widely known are traded readily even to "outsiders" as a way to express welcome and openness; however, valuable papers showing more productive and esoteric tows would not be traded to these fishermen. Bob, as a non-native to Plymouth, was in the position of the "outsider" and was not given valuable papers until he had worked from Plymouth for several years. He was fortunate to have information from other sources which allowed him to fish the area until trust and conditions of reciprocity were established. Bob now possesses many papers from several individual fishermen from within this community.

Figure 21.1 shows a typical paper provided by Bob for an area within a day trip of Plymouth. This figure shows the specificity of the pathways around and through rocky areas and "humps." The multiple lines on the paper represent individual sets of a boat's net and where it was towed during each set. This paper was used repeatedly, probably over the course of several years. Some pathways are more often traveled than others; it can be assumed that they were more successful tows in terms of fish caught. Some pathways may represent unsuccessful trials to maneuver through rocks or humps. The paper in Figure 21.1 is typical of the papers belonging to trawl fishermen who work inshore. Clearly there is a complex and detailed "landscape" that is known to varying degrees by different fishermen in this community. To successfully fish in the "territory" of the Plymouth fishermen, it was necessary for Bob to enter into this system of exchange of environmental information and thereby produce along with the other fishermen of Plymouth a community and shared commons.

The community of fishermen in Plymouth is an ongoing process rather than a fixed and traditional entity (Gudeman and Rivera Gutiérrez 2002; Ratner and Rivera Gutiérrez 2004). It is neither permanent nor closed and changes as new fishermen try to enter and others exit. Also, the commons itself, the space produced and maintained by Plymouth fishermen through processes that clearly limit access, remains unbounded insofar as it cannot act as the community's traditional or exclusive territory. These processes and practices introduce an ambiguity relative to Bob's identity and representations of the commons. While he is certainly mobile and independent (e.g., he moves between several such communities and locations in search of

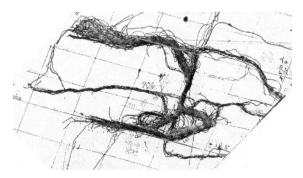

Figure 21.1 A typical plotter produced paper for a mixed (rocky with pathways bottom).

better opportunities), he is, nevertheless, dependent upon community and community-produced knowledge for his success. The locations into which he ventures are similarly ambiguous; they are "open access," yet successful utilization of resources is only attained through a negotiation with other fishermen. It would seem that it is possible to read the fishermen of Plymouth as both independent utility-maximizing individuals on an open access resource and as community members operating within territories of limited access.

Conclusion

The fishing commons of New England is represented by a dominant neo-classical discourse of fisheries as a site of potential tragedy only redeemable through a movement toward enclosure and privatization of access to fisheries resources. While this particular narrative of the commons (in fisheries and elsewhere) has been roundly criticized and qualified, it remains hegemonic in New England and is increasingly used to represent fisheries throughout the world. The pervasiveness of this representation is due not only to its enticing promise of delivering stability and environmental sustainability but also to the impossibility of any alternative. This and other representations of the commons relegate economic difference to an epoch before (or beyond) the present of the capitalist commons.

Rereading the commons, however, as a site of multiple economic identities attempts to displace the binary of pre-capitalist past and capitalist present. In the case of Bob and the fishermen of Plymouth, past and present are seen to coexist and mingle in unexpected ways. The dominant capitalo-centric representation of the commons fixes the economic identity of Bob such that his participation in a community economy would be unimaginable or, worse, a sign of schizophrenia. Where the economy is seen as diverse, the multiple (and mutable) nature of Bob's economic subjectivity allows us to imagine and, perhaps, facilitate more than one economic future (and economic past) for the fishermen of Plymouth.

Acknowledgments

I would like to thank Paul Robbins and Nik Heynen for their valuable comments on the original version of this chapter which appeared as St. Martin, K. (2005) "Disrupting Enclosure in New England Fisheries," *Capitalism, Nature, Socialism* 16(1): 63–80.

Notes

1 The interviews were part of an oral history project (S-K Grant 96-NER-166) belonging to the Gloucester Fishermen's Wives Association (GFWA). Angela Sanfilippo (director of the GFWA), Dr. Madeleine Hall-Arber (MIT, Sea-Grant Program), and Dr. Christopher Dyer (URI, Marine Affairs) were the principal

investigators on the project. The author participated in the research design, interviews, and subsequent analysis.
2 Gibson-Graham (1996, p. 87) is here referring to the work of M. Moon, E. K. Sedgwick, B. Gianni, and S. Weir, "Queers in (single family) space," *Assemblage*, 24, 1994, pp. 30–7.

References

Acheson, J. M. (1988) *The Lobster Gangs of Maine*, Hanover, NH: University of New England Press.
Apostle, R., McCay, B. and Mikalsen, K. H. (2002) *Enclosing the Commons: Individual Transferable Quotas in the Nova Scotia Fishery*, St. John's, Newfoundland: Institute of Social and Economic Research.
Callari, A. (2004) "Economics and the Postcolonial Other," in E. O. Zein-Elabdin and S. Charusheela (eds.) *Postcolonialism Meets Economics*, New York: Routledge, pp. 113–29.
Community Economies Collective (2001) "Imagining and Enacting Noncapitalist Futures," *Socialist Review* 28(3+4): 93–135.
Davis, A. (1996) "Barbed Wire and Bandwagons: A Comment on ITQ Fisheries Management," *Reviews in Fish Biology and Fisheries* 6: 97–107.
Doeringer, P.B., Moss, P.I. and Terkla, D.G. (1986) *The New England Fishing Economy: Jobs, Income, and Kinship*. Amherst, MA: University of Massachusetts Press.
Eckert, R. D. (1979) *The Enclosure of Ocean Resources: Economics and the Law of the Sea*. Stanford, CA: Hoover Institution Press.
Garcia, S. M., Cochrane, K., van Santen, G. and Christy F. (1999) "Towards Sustainable Fisheries: A Strategy for FAO and the World Bank," *Ocean and Coastal Management*. 42: 369–98.
Gibson-Graham, J. K. (1996) *The End of Capitalism (As We Knew It): A Feminist Critique of Political Economy*, Oxford: Basil Blackwell.
—— (2006) *A Postcapitalist Politics*, Minneapolis, MN: University of Minnesota Press.
Gibson-Graham, J. K., Resnick, S. A. and Wolff, R. D. (2000) *Class and Its Others*, Minneapolis. MN: University of Minnesota Press.
Gibson-Graham, J. K. and Ruccio, D. (2001) "'After' Development: Re-imagining Economy and Class," in J. K. Gibson-Graham, S. Resnick and R. Wolff (eds.) *Re/presenting Class*, Durham, NC: Duke University Press, pp. 158–81.
Gudeman, S. and RiveraGutiérrez, A. (2002) "Neither Duck Nor Rabbit: Sustainability, Political Economy, and the Dialectics of Economy," in J. Chase (ed.) *The Spaces of Neoliberalism: Land, Place and Family in Latin America*, Bloomfield, CT: Kumarian Press, Inc., pp. 159–86.
Hannesson, R. (2004) *The Privatization of the Oceans*, Cambridge, MA: The MIT Press.
Hardin, G. (1968) "The Tragedy of the Commons," *Science* 162: 1243–8.
—— (1977) "Denial and Disguise," in G. Hardin and J. Baden (eds.) *Managing the Commons*, San Francisco: W.H. Freeman, pp. 45–52.
Leyshon, A., Lee, R. and Williams, C. (2003) *Alternative Economic Spaces: Rethinking the "Economic" in Economic Geography*, London: Routledge.
McCarthy, J. (2005) "Commons as Counterhegemonic Project," *Capitalism, Nature, Socialism*. 16(1): 9–24.

McCay, B. J. and Brandt, S. (2001) "Changes in Fleet Capacity and Ownership of Harvesting Rights in the United States Surf Clam and Ocean Quahog Fishery," in R. Shotton (ed.) *Case Studies on the Effects of Transferable Fishing Rights on Fleet Capacity and Concentration of Quota Ownership*, Rome: Food and Agricultural Organization of the United Nations.

Mansfield, B. (2004) "Neoliberalism in the Oceans: 'Rationalization,' Property Rights, and the Commons Question," *Geoforum* 35(3): 313–26.

Marx, K. (1976) *Capital*, vol. 1, London: Penguin Books.

Ostrom, E, Gardner, R. and Walker, J. (1994) *Rules, Games, and Common-Pool Resources*, Ann Arbor, MI: The University of Michigan Press.

Palsson, G. (1994) "Enskillment at Sea," *Man* 29(December): 875–900.

Palsson, G. and Helgason, R. (1995) "Figuring Fish and Measuring Men: The Quota System in the Icelandic Cod Fishery," *Ocean and Coastal Management* 28(1–3): 117–46.

Pavlovskaya, M. (2004) "Other Transitions: Multiple Economies of Moscow Households in the 1990s," *Annals of the Association of American Geographers* 94(2): 329–51.

Ratner, B. D. and Rivera Gutiérrez, A. (2004) "Reasserting Community: The Social Challenge of Wastewater Management in Panajachel, Guatemala," *Human Organization* 63(1): 47–56.

Robbins, P. (2004) *Political Ecology: A Critical Introduction*, Malden, MA: Blackwell.

Schlager, E. and Ostrom, E. (1992) "Property-Rights Regimes and Natural Resources: A Conceptual Analysis," *Land Economics* 68(3): 249–62.

St. Martin, K. (2001) "Making Space for Community Resource Management in Fisheries," *Annals of the Association of American Geographers* 91(1): 122–42.

—— (2005) "Mapping Economic Diversity in the First World: The Case of Fisheries," *Environment and Planning A* 37: 959–79.

Wilson, J. A. (1990) "Fishing for Knowledge," *Land Economics* 66(1): 12–29.

Part IV

Commentary

22 Researching resistance in a time of neoliberal entanglements

Juanita Sundberg

The chapters by David Correia, Wendy Wolford, and Kevin St. Martin pose a challenge to "capitalocentric" narratives about economic and cultural processes, in which market relations inevitably triumph (see St. Martin's chapter). Focusing on sites as diverse as New Mexico, Brazil, and New England, the researchers refuse the linearity and teleology of capitalocentrism by demonstrating the ways in which differing socio-economic relations overlap in time and space as a result of continuity in livelihood practices, opposition movements, and surprising convergences. Despite the best intentions of powerful elites to create the discursive and legislative conditions necessary to privatize natural resources and enroll individuals into the market economy as wage laborers, people resist, policies go awry, and contradictions emerge.

Correia's analysis of "'Continuous and Ample Supply': Sustained Yield Timber Production in Northern New Mexico" documents one such instance in the late 1940s. In response to a number of perceived environmental problems (depleted grazing lands; crises in industrial timber production), the Forest Service instituted a set of policies to increase the role of forestry in the local economy. Local residents, whose collective identity centered on historical rights to land and livelihood, confounded such efforts by employing a strategy common to contemporary resistance movements the world over: they called upon government officials to enforce existing laws ensuring worker protections. By attending to the "specific historical, geographical and ecological conditions of production in New Mexico," Correia finds that the regulatory framework "impaired the operation of the very markets in nature and labor [it] sought to establish."

Wolford shows that the Rural Landless Workers (MST) movement deploys similar resistance strategies by using Brazilian law as a launching pad for moral and discursive contests. With their motto "land for those who work it," the MST holds the government accountable to the constitution, which creates the moral and legislative conditions for land re-distribution. Moreover, the MST contests elite discourses and practices favoring the "invisible" hand of the market by pointing to the ways in which the state makes access to the market *unequal* through specific policies and incentives.

Wolford's analysis of the MST leadership's narratives allows her to illustrate how the movement has been able to posit alternative conceptions of land as "the key to a meaningful livelihood and to effective citizenship in Brazil."

St. Martin's chapter highlights the limitations of capitalocentric narratives about common property regimes, arguing that they mask alternative solutions to resource crises. Drawing on his qualitative research with New England fishers, St. Martin reveals the "community relations, community-economic processes, territorializations, and local understandings and representations of resources" that constitute fisheries commons. In showing the ways these otherwise invisible sets of relations articulate with multiple economic/class processes and identities, St. Martin "opens the space of the commons to more than one (i.e. capitalist) future."

In their analysis of resistance to the privatization of natural resources, Correia, Wolford, and St. Martin employ research methodologies that are attentive to the everyday narratives of individuals and collectives involved in enacting such struggles. The authors consider the richness of lived experiences in time and place and show how individuals and organizations enter into struggles over differing visions of the future. Bringing the particularity of situated knowledges and practices to life precludes the production of overgeneralizing narratives about neoliberal projects.

Moreover, by documenting what resistant movement participants are saying and how they frame their critiques, the researchers create space for the articulation of alterative conceptualizations of land and livelihood. In so doing, the authors ensure that movements are not authored or authorized by researchers, but by those individuals and collectives whose lives depend upon the outcome of their struggles.

That being said, I am uncomfortable delineating a clear boundary between researchers and activists, for we are all subjected to and implicated in neoliberal projects. Despite explicit critiques of neoliberal regulation of nature, critical scholars in privileged positions (here, I include myself) benefit directly from these processes in terms of the foods we eat, the beverages we drink, the clothes we wear, etc. We cannot analyze or critique our way out of the socio-economic relations in which we are embedded and from which we benefit. In this time of neoliberal *entanglements*, we must ask if our research methodologies enable us to produce knowledge that reveals the ways in which we are intertwined in such processes.

Ultimately, this question speaks to broader questions about our roles as critical scholars. In other words, how are *our* politics to be made manifest? Is the study of resistance movements a political act in and of itself? Or do our entanglements oblige us to take a more explicit stance regarding the when, where, how and why of our own political engagements, agendas and practices?

One barrier to critically engaged research is established boundaries between producers and objects of knowledge, academics and activists. Scholarly conventions framing certain populations as appropriate objects of study are rooted in colonial and imperial relations of power between specific geopolitical units and the social groups within those units. Like it or not, most social scientists are trained in research traditions touched by Area Studies, which is built upon the assumption that collecting knowledge about geopolitical others is in the interests of hegemonic actors (Mato 2000; Morris-Suzuki 2000). To resist reproducing the distance and inequality at the heart of our research traditions requires self-reflexively subjecting our methodologies to radical critique (Maxey 1999).

As a means of decolonizing research, Daniel Mato (2000) suggests moving away from doing studies *on* specific populations to *studying with* those groups. There are many models for *studying with*, ranging from the use of specific methods to produce more inclusive accounts that emphasize participants' ways of understanding, to participatory "methodologies and epistemologies that aim to effect change for and with research participants" (Pain and Francis 2003: 46). One primary goal of such approaches is to challenge established hierarchies by framing objects of study as speaking subjects, thereby emphasizing that the production of knowledge is an inter-subjective process, the outcome of embodied interactions between individuals who are differently located, yet entangled (see Gibson Graham 1994; Nagar 2002). A second goal is to make research into a mechanism for direct social action and change (Pain and Francis 2003).

Studying with, however, may not be a suitable approach for all research projects. Moreover, there is no guarantee that *studying with* will reveal the extent to which researchers are entangled with research subjects in asymmetrical relations of power constituted by and reflected in global and regional neoliberal institutions and regulatory regimes.

In thinking about how to do research that is open to and informed by neoliberal entanglements, my mind wanders back to an evening in Vancouver when a group of people got together to read parts of the *Sixth Declaration of the Selva Lacandona* (2005) and talk about what it meant for us.[1] The *Sixth Declaration* invites people who are engaged in struggles against neoliberalism and for democracy, liberty and justice to *walk with* the Zapatista Army of National Liberation. The Zapatista's vision of *walking with* breaks with previous notions of solidarity and leftist political alliances. To *walk with* the Zapatistas means to be involved in the struggle for a just world from and in our own jungle, whether that is Vancouver, Canada or Austin, Texas. *Walking with* does not mean helping the Zapatistas nor does it mean being like the Zapatistas and following their orders. *Walking with* is a new mode of doing politics locally and globally.

What would it be like to frame research in terms of *walking with* differently situated others in intersecting, yet distinct and unequally constituted struggles? In thinking about *walking with* as a metaphor for research, I immediately envision collaborations. I am inspired, for example, by Geraldine Pratt's collaborative relationship with the Philippine Women Centre to document the experiences of Filipina women working in Canada through the Live-in Caregiver Program (Pratt 2004). As Pratt (2002: 198) describes it, the "relationship is not based on an identity or a shared set of experiences." Rather, common ground was forged from shared analyses of political economy, norms about social justice, and a critique of the hypocrisy in countries like Canada, wherein universal ideals of citizenship and equality are unevenly applied. We also might consider Cindi Katz's (2001) notion of "counter-topographies" as a model. Counter-topographies of neoliberalism might analyze how people – as individuals or collectives – in our own community are connected to people in another community by over-arching neoliberal policies, while also showing how such policies are informed by the specifics of fully contextualized places.

Clearly, there will be as many ways of *walking with* as there are individuals and communities. In my view, however, the notion of *walking with* presupposes that scholars are explicit about their political position in relation to their research and writing practices. By revealing our own entanglements in neoliberal projects, we break down the relations of hierarchy and distance that constitute conventional research practices and make academic knowledge production a key *site* and *means* of resistance.

Note

1 The *Sixth Declaration of the Selva Lacandona* by the Zapatista Army of National Liberation is available online. I accessed the document at www.indymedia.org/en/2005/07/117697.shtml in June 2006.

References

Gibson-Graham, J-K. (1994) "'Stuffed if I Know!': Reflections on Postmodern Feminist Social Research," *Gender, Place and Culture* 1: 205–24.
Katz, C. (2001) "On the Grounds of Globalization: A Topography for Feminist Political Engagement," *Signs* 26: 1213–34.
Mato, D. (2000) "Not 'Studying the Subaltern,' but Studying with 'Subaltern' Social Groups, or, at Least, Studying the Hegemonic Articulations of Power," *Nepantla: Views from South* 1(3): 479–502.
Maxey, I. (1999) "Beyond Boundaries? Activism, Academia, Reflexivity and Research," *Area* 31(3): 199–208.
Morris-Suzuki, T. (2000) "Anti-Area Studies," *Communal/Plural* 8: 9–23.
Nagar, R. (2002) "Footloose Researchers, 'Traveling' Theories, and the Politics of Transnational Feminist Praxis," *Gender, Place and Culture* 9(2): 179–86.
Pain, R. and Francis, P. (2003) "Reflections on Participatory Research," *Area* 35(1): 46–54.
Pratt, G. (2002) "Collaborating across Our Differences," *Gender, Place and Culture* 9(2): 195–200.
—— (2004) *Working Feminism*, Philadelphia, PA: Temple University Press.

23 What might resistance to neoliberalism consist of?

Michael Watts

In a society which is to preserve freedom of choice of the consumer and free choice of occupation, central direction of all economic activity presents a task which cannot be rationally solved under the complex conditions of modern life.
(Friedrich Hayek, *Collectivist Economic Planning*)

Karl Polanyi's *The Great Transformation* and Friedrich Hayek's *The Road to Serfdom* were both published in 1944. Hayek, an Austrian economist trained at the feet of Ludwig von Mises but forever associated with a largely non-economic corpus produced at the London School of Economics and the universities of Chicago and Freiburg between 1940 and 1980, is widely recognized as one of the leading intellectual architects of the neoliberal counter-revolution. Margaret Thatcher pronounced that "this is what we believe" as she slammed a copy of Hayek's *The Constitution of Liberty* onto the table at Number 10 Downing Street during a Tory Cabinet meeting. Hayek's critique of socialism – that it destroys morals, personal freedom and responsibility, impedes the production of wealth and sooner or later leads to totalitarianism – is the *ur*text for market utopians. Collectivism was by definition a *made* rather than a *grown* order: it was, Hayek said, constructivist rather than evolutionary, organized not spontaneous, a "taxis" (a made order) rather than a "cosmos" (a spontaneous order), an economy rather than a "catallaxy," coerced and concrete rather than free and abstract (see Gamble 1996: 31–2). Its fatal conceit was that socialism (and social democracy for that matter) admitted the "reckless trespass of *taxis* onto the proper ground of *cosmos*" (Anderson 2005: 16).

The other half of Hayek's project was a robust defense of western civilization – that is to say of liberty, science and the spontaneous orders that co-evolved to form modern society ("Great Society" as he termed it). It was a buttressing of the liberal (unplanned) market order from which the preconditions of civilization – competition and experimentation – had emerged. Hayek, like Weber, saw this world as an iron cage constituted by impersonality, a loss of community, individualism and personal responsibility. But (unlike Weber) Hayek saw these structures, properly understood, as expressions of liberty. From the vantage point of the 1940s this (classical) liberal project was, as Hayek saw it, under threat; what passed as liberalism was a travesty, a diluted and distorted body of ideas corrupted by con-structivist rationalism (as opposed to what he called "evolutionary ration-alism"). The ground between liberalism and much of what passed as Keynesianism or social democracy was, on the Hayekian account, cata-strophically slight. What was required, as he made clear at the founding of the Mont Pelerin Society in 1947, was a restoration, a purging of true lib-eralism (the removal of "accretions"). There was to be no compromise with collectivism; the seized territory had to be regained. In his writing and his promotion of think-tanks like the Institute of Economic Affairs in Britain – the brains trust for the likes of Keith Joseph and Margaret Thatcher – Hayek aggressively launched a cold war of ideas. He was part of the quartet of European theorists (Carl Schmitt, Leo Strauss and Michael Oakeshott were the others) whose ideas, while standing in a tense relationship to one another, came to shape a large swath of the intellectual landscape of the early twenty-first century (see Anderson 2005). Hayek was neither a simple conservative or libertarian, nor a voice for *laissez-faire* ("false rationalism" as he saw it). He identified himself with the individualist tradition of Hume, Smith, Burke and Menger, thereby providing a bridge that linked his short-term allies (conservatives and libertarians) to classical liberals in order to make common cause against collectivism (Gamble 1996: 101). To roll back the incursions of taxis required a redesign of the state. A powerful chamber was to serve guardian of the rule of law (striking all under 45 years off the voting roll), protecting the law of liberty from the logic of popular sover-eignty. As Anderson (2005: 17) notes, the correct Hayekian formula was "demarchy without democracy."

Polanyi was a Hungarian economic historian and socialist who believed that the nineteenth-century liberal order had died, never to be revived. By 1940, "every vestige" of the international liberal order had disappeared, the product of the necessary adoption of measures designed to hold off the ravages of the self-regulating market (market despotism). It was the conflict between the market and the elementary requirements of an organized social life that made some form of collectivism or planning inevitable (1944). The liberal market order was, *contra* Hayek, not "spontaneous" but a planned development, and its demise was the product of the market order itself. A market order could just as well produce the freedom to exploit as it could the freedom of association. The grave danger, in Polanyi's view, was that liberal utopianism might return in the idea of freedom as nothing more than the advocacy of free enterprise, the notion that planning is nothing more than "the denial of freedom" and that the justice and liberty offered by regulation or control becomes nothing more than "a camouflage of slavery" (1944: 258). Liberalism in this account will always degenerate, ultimately compromised by an authoritarianism that will be invoked as a counter-weight to the threat of mass democracy. Modern capitalism con-tained the famous "double movement" in which markets were serially and coextensively disembedded from, and re-embedded in, social institutions and relations – what Polanyi called the "discovery of society." In particular, the possibility of a counter-hegemony to the self-regulating market could be found in resistance to the commodification of the three fictitious commodities (land, labor and money); such reactions represented the spontaneous defense of society (Burawoy 2003).

The market and anti-market mentalities of Hayek and Polanyi were both forged in the context of fascism, global economic depression, revolution and world war. To look back on the birth of *The Great Transformation* and *The Road to Serfdom* from the perch of 2006 (and the three chapters on offer here) is quite salutary: we see American empire (military neoliberalism), a global war on terror, the dominance of unfettered global finance capital, a world-wide Muslim resurgence, a phalanx of "failed states" (otherwise known as the failure of secular nationalist development) and a raft of so-called anti-globalization movements, and the rise of civic regulation. There has been a consequent Polanyi boom (see C. Williams 2005) within the academy, but fewer careful readings of the Hayekian ideas that helped spawn these developments. From within the bowels of this turmoil, the Hayekian vision is triumphant – the Liberal International has come to pass. Its long march, from Mont Pelerin to the collapse of the Berlin Wall and TINA ('There Is No Alternative') took about 40 years and, according to Harvey's (2005) brief history, passed through the Chicago Boys in Chile, the IMF/IBRD complex, the Reagan–Thatcher revolutions and the corporate (class) seizure of power in the 1970s against a backdrop of declining profitability and income share. Even if "global neoliberalism" has now assumed a neo-conservative and military cast (see Saad-Filho and Johnson 2005), nobody seems to question its hegemony: as Gramsci might have put it, there has been a Hayekian "passive revolution" from above (see Arrighi and Silver 2003). We have witnessed what the Left's great pessimist Perry Anderson (2000) has dubbed a "neoliberal grand slam," with neoliberalism ruling undivided across the globe as the most successful ideology in world history (2002). This "fluent vision" of the Right has no equivalent on the Left: Anderson cedes that embedded liberalism (let alone something called socialism) is now as remote as "Arian bishops," resistances are like "chafe in the wind," and the Left can only "shelter under the skies of infinite justice." From what sources, then, are counter-hegemonic responses to appear, and what might resistance to neoliberalism possibly consist of at this point?

The first thing that needs to be said is that the very process by which neoliberal hegemony was established – and against which forms of resistance are to be assessed – remains a story for which at present we have no full genealogy. The cast of characters may be lined up – from the school of Austrian economics to the Reagan–Thatcher–Kohl troika – but this explains very little. Neoliberalism can be seen as a class reaction to the crisis of the 1970s (as both Milton Friedmann and Bob Brenner concur). The global multilaterals and the Wall Street Treasury certainly imposed brutal forms of economic discipline – structural adjustment – to eradicate forever any residue of collectivism in the Third World. But beyond these general descriptions we are left with paradoxes and questions, of which I will name just a few. Why did the LSE and Chicago – the centers of Fabianism and a certain sort of American liberalism – become the forcing houses of neoliberalism? Hayek, after all, was not associated with the Economics Department; it was the arrival of Ronald Coase at Chicago that marked a neoliberal turning point. How did the World Bank – a bastion of postwar development economics and a certain sort of statism – become the voice of *laissez-faire*? Harry Johnson (who held Chairs at the LSE and Chicago) certainly figures in the process, but how can we explain economic liberalism's capture of key sectors of the Bank (often by second-rate economists) against a backdrop of robust Keynesianism? How did the ideas of Peter Bauer and Deepak Lal (who in a 1983 IEA paper declared that development economics was dead) gain traction? Criticism of Keynes dovetailed with the anti-statism leveled by many on the Left during the 1970s. In other words, tracing the ways in which government failures came to outweigh market failures in development thinking demands a complex picture of discursive contestations and political practices. Indeed, by the mid to late 1970s, many of neoliberalism's intellectual architects (Friedmann among them) claimed that nobody took their ideas seriously – Hayek believed *The Road to Serfdom* had ruined his career and marginalized his entire project. It was the inflation of the 1970s, said Friedmann, that revealed the cracks within the Keynesian edifice. The point is that the "neoliberal grand slam" was preceded by decades of pretty mediocre hitting, pessimism, and contestation, and that the class forces around and through which embedded liberalism had been built necessarily shaped the manner and forms in which the counter-revolution could proceed (if at all). How to think about resistance to neoliberalism, then, turns on how one sees this long march through institutions.

The three chapters by Wolford, Correia and St. Martin speak directly to the incompletion and unevenness of neoliberalism. They retain a strongly Polanyian cast (land (and sea), labor and the politics of fictitious commodities figure centrally) and local points of reference. In Brazil, the differing forms of land reform represent, as one would expect, differing configurations of class forces struggling over core elements of the neoliberal project. The New Mexico sustained yield timber unit unleashed a firestorm of local struggles, over labor conditions and over the legitimacy of local versus non-local contractors. In the New England fisheries, an orthodox narrative of the privatization of the commons reveals dependence upon another commons – shared knowledge – which belies any simple sense of class identity or indeed of primitive accumulation (the binary of pre-capitalist relics versus emergent capital). Each of these chapters stands, however, in a complex relation to the realities of neoliberalism. In some cases, it is not clear how the term is being used other than to imply a sort of privatization or deregulation. In all cases, what passes as neoliberalism is encased with a carapace of (re)regulatory and coordinating institutions that extend beyond the market narrowly construed.

For these reasons, all the chapters can be pushed further, I think. In Wolford's case from Brazil, the invocation of "community" as a counter-force to market-led reform fails to come to terms with the extent to which the community has become the neoliberal form *par excellence* of modern governmentality. At the very least, the double movement of MST versus MLAR must be placed upon a larger landscape of the class forces shaping the complicated trajectories of each. In St. Martin's New England study, the invocation of shared knowledge as a counter-weight to the binary logic of the commons seems to me to be a weak reed to carry the ambitious claims about alternative economies. After all, commons of this sort are quite constitutive of modern industrial districts and the conventions central to much industrial innovation. In other words, "the commons" understood in this sense can be compatible with some quite conventional political-economic projects. In Correia's case, the struggles over New Mexico's forests seemed to produce a stalemate at best, one that left the floor open to big timber companies. Assessing whether resistance could have produced other outcomes seems to me requires a more wide-ranging analysis, of the sort conducted by Kosek (2006), which draws into the story questions of race, ethnicity and historical memory that are only gestured to in the chapter.

In sum, these chapters all explore examples of local spontaneous defenses of society. If we are to map the larger landscape, resistance to neoliberalism must, as Polanyi noted, take account of divergent national responses and global reactions. Here there is perhaps reason to be less gloomy than Anderson. In contrast to the 1990s prospect of globalization triumphalism, today the WTO is something of a shambles, and ferocious fights within and between the IMF and the IBRD suggest that the neoliberal project has confronted resistances along many fronts. Whether the "multitude" or the Movement of Movements currently represents a serious Polanyian "global double movement" is perhaps an open question. That anti-imperialist political Islam does is surely incontestable (RETORT 2005). The national level, in turn, reveals a vast unevenness in the nature of the liberal assault and the resistances to it. The recent students' and minorities' revolts in France, worker takeovers in Argentina, oil nationalization in Bolivia, and insurgencies in Iraq are all of a piece in the sense being reactions against neoliberalism, even as their national dynamics are massively at a variance to one another.

It is from the irreducible centrality of class, exploitation and the contradictory reproduction of capitalism as a dynamic and changing system that resistance and alternatives can emerge. Fong and Wright's (2004) work on "empowered participatory governance" has the very great strength of linking participatory budgeting in Brazil with decentralized planning in India with school reform in Chicago (see also M. Williams 2005). Such "vigorous underground experimentalism" (Unger 2000) may represent a counter-weight to the Hayekian revolution in its multiplicity of forms. But we should also reflect upon Andrew Gamble's concluding observation in his exemplary book on Hayek himself, namely, that Hayek has most to offer socialists. His analysis of knowledge, coordination and institutions revealed that the most effective forms of social organization were decentralized and democratic. Hayek's own elitism and classical liberalism predisposed him to a sort of political despotism, but it is perhaps the traditional Left goals of solidarity and equality that might most benefit from Hayek's ruminations on dispersed knowledge, coordination, and spontaneous orders.

References

Anderson, P. (2000) "Renewals," *New Left Review* 1: 5–24.
—— (2002) "Internationalism: A Breviary," *New Left Review* 14: 5–25.
—— (2005) *Spectrum.* London: Verso.
Arrighi, G. and Silver, B. (2003) "Polanyi's Double Movement," *Politics and Society* 31/2: 162.
Burawoy, M. (2003) "For a Sociological Marxism," *Politics and Society* 31/2: 162–83, available at: www.sociology.berkeley.edu/faculty/BURAWOY/
Fong, A. and Wright, E. (eds.) (2004) *Deepening Democracy*, London: Verso.
Gamble, A. (1996) *Hayek: The Iron Cage of Liberty*, Boulder, CO: Westview Press.
Harvey, D. (2005) *A Brief History of Neoliberalism*, Oxford: Clarendon Press.
Kosek, J. (2006) *Understories*, Durham, NC: Duke University Press.
RETORT (2005) *Afflicted Powers*, London: Verso.
Saad-Filho, A. and Johnson, D. (2005) *Neoliberalism: A Critical Reader*, London: Pluto Press.
Unger, R. (2000) *Democracy Realized*, London: Verso.
Williams, C. (2005) *A Commodified World*, London: Zed Press.
Williams, M. (2005) "Democratic Communists: Party and Class in South Africa and Kerala, India," PhD dissertation, Department of Sociology, University of California, Berkeley.

Part V

Conclusion

24 Neoliberal ecologies

Noel Castree

Imagine the impossible: a world without geography. This would be a head-of-a-pin world where everything takes place at one, and only one, site. Or, to view things another way, it would be an isotropic planet where all sorts of different places, peoples and environments were entirely homogenous – and thus not different at all. Now imagine how this world would look if opposed theoretical perspectives on neoliberalism were to be made flesh. Friedrich Hayek and Karl Polanyi, as Michael Watts (Chapter 23, this volume) reminds us, published their *magnum opuses* in the same year (1944), the beginning of a period when classical (or 'disembedded') liberalism became yesterday's news. These days, *The Road to Serfdom* and *The Great Transformation* are cited by champions and critics of neoliberalism respectively as making among the strongest cases for and against the phenomena. So how, exactly, would our ageographical world look if the arguments of Hayek and Polanyi were to hold true on the ground?

The Hayekian vision, of course, is a rose-tinted one. All or most human transactions would be market-based or market-like; the state would be a facilitator and referee but little more; individuals and communities would enjoy maximum liberty; an ethic of self-help and personal responsibility would be universal – in short, freedom would rule the world and all social choices would be 'optimal' by virtue of being market-governed within definite resource constraints. Polanyi, of course, saw things the other way around. For him, the expansion of the market would displace other valued means of living and working; the putatively 'non-interventionist' state would, in his view, permit the social and ecological abuses associated with market rule to proliferate in time and space; the economic and juridical freedoms that Hayek championed would, for Polanyi, be partial and ultimately illusory; and, finally, Polanyi argued that the creation of asocial, atomised individuals and communities would be challenged by organised resistance resulting from the general lack of social and environmental protections – what he famously called 'the double movement'. Here, then, we have two opposed views of neoliberalism writ large: a utopian and a dystopian one which, if we imagine an isotropic world, would find practical expression in homogenous geographies of happiness and strife respectively.

Neither Hayek nor Polanyi were simply 'theorists', of course. Both understood the messiness of history and geography well enough. But today their plenary arguments are used as heavy weapons in the often polarised debates between neoliberals and their left-wing (sometimes right-wing) critics. The tone of these debates is often combative and universal: blanket arguments 'for' and 'against' abound in academic, policy and public circles. 'Third way' arguments mediate the conflict, but often in equally grand and generalising terms. As the chapters in this volume show so well, geographers have an important and distinctive contribution to make to societal understandings of neoliberalism, surely one of the defining phenomena of our time. Typically, we eschew the grand claims emanating from those who work, implicitly, with an ageographical imagination. Instead, we pay close attention to the dialectic of connectivity and difference, similarity and particularity that is the central feature of contemporary life: we call it uneven development.[1] This does not, however, make us mere merchants of fact. Despite this 'gazeteer' view of academic geography still prevailing in large sections of academia and the public realm, geographical analysis today typically troubles the theory–empirics dualism. It does so – and again, the chapters assembled here testify to this powerfully – by showing how 'process' does not simply produce empirical 'outcomes' but is itself partly *produced* by its forms of 'expression'. Here, then, the suggestion that there are 'general' processes that *precede* their merely 'contingent' forms of spatio-temporal realisation is called into question. This does not mean that the neoliberal world is bereft of common logics or consequences across time and space. But it does mean that the neoliberal project is shown to be *constitutively differentiated* at some level – and that this differentiation might make a difference to how we understand and evaluate those things the grand abstraction 'neoliberalism' inadequately signifies.

The value of this for all those interested in neoliberalism – whatever their political persuasion and regardless of whether or not they are geographers – is plain to see. The particular shibboleths about neoliberalism that analysts of different political and disciplinary persuasions seem to hold dear cannot usually withstand the rigours of theoretically informed but empirically grounded geographical research. This kind of research keeps all of us who are interested in the neoliberal project honest, both cognitively and normatively. It allows us to see what 'neoliberalism' actually *is* – a set of differentiated yet often connected neoliberalisations – rather than what ideal-typical arguments say it *will* or *should be*. Clearly, this has important implications for both understanding and evaluation, as well as policy prescription.

In the case of this volume, geographers' research into neoliberalisations (in the plural) has a particular originality over and above the general strengths just noted. As the editors observe in their overviews, the connections between neoliberal practices and the natural environment are only just beginning to be fathomed. The authors whose previously published work is abridged here are among the first in the social and environmental sciences to offer in-depth analyses of the neoliberalisation of nature in the modern world. These authors (myself included) would happily describe themselves as 'critical geographers' (or, if not, be so described by their peers). Their research, as readers of this book will hopefully have realised, has at least five strengths. First, it spans a wide range of resources and environments: from forests to wetlands to deer farms and beyond. Second, it attends variously to all the salient dimensions of nature's neoliberalisation: namely, logics (who is driving it and why?); processes (how does it actually operate?); outcomes (what occurs, who or what is affected, and how?); and evaluations (how should we judge it?). Third, the research often pays close attention to the materiality of the non-human world: nature in its various forms figures as a biophysical actor not a *tabula rasa* or neutral 'backdrop'. This is important because nature can be shown to alter the workings and outcomes of neoliberal governance ideas, rules and mechanisms. Fourth, close attention is also often paid to issues of scale-crossing and scale-jumping: the links between different socially constituted geographical scales in terms of logics, processes and outcomes are often strongly accented, so that one or other scale of environmental governance is not hypostatised or fixated upon as if others can be conveniently bracketed out.[2] Finally, critical geographical research into nature's neoliberalisation covers a remarkable array of places, regions and countries. All of this, potentially at least, offers readers of the literature and future researchers of the topic a fairly comprehensive sense of why and how neoliberal environmental governance operates today, with what effects and with what actual and desirable normative responses.

We can summarise these strengths with recourse to the label 'neoliberal ecologies' (my title).[3] Traditionally, the term 'ecology' denoted a relatively stable biotic complex: an ensemble of more-or-less intricate relations specific to a place, region or larger territorial entity. The term neoliberal ecologies expands this meaning to denote diverse but often interlinked and possibly dynamic *biosocial complexes* in which neoliberal policies and practices remake (and are remade or resisted by) the non-human world. These biosocial complexes are often surprising as they take shape and stabilise, however precariously.

As I just noted, all the authors writing in this book can be labelled critical geographers. Yet this does not commit them to an unremittingly negative – that is to say, dogmatic – view of the effects of neoliberal policies on natural resources or the natural environment more generally. For instance, Bakker's chapter on water concedes that real environmental gains have been made in English and Welsh water quality since utilities privatisation in 1989. Yet, as Wendy Larner notes in her commentary, even when their researches indicate otherwise, critical geographers studying nature's neoliberalisation do not *always* arrive at the ambivalent conclusions their analyses typically demand. We are, perhaps, still saddled with our own shibboleths – at least at the normative level – and have some way to go before we abandon them for the more supple understandings geographical inquiry can deliver.

Having concluded my so-far positive assessment of this book's chapters with this mildly critical observation, let me offer a few programmatic suggestions in the remainder of my commentary. What remains to be done if we are to properly understand, evaluate and intervene in neoliberal ecologies? Clearly, this is partly an academic question, but also partly a question of policy and public understanding. So let me take each element in turn. In purely research terms, it seems to me that a close reading of the literature assembled in this volume reveals the following unresolved issues. First, there remains much work to be done comparing and contrasting the sort of single-site or single-resource studies found in this volume. Readers of *Neoliberal Environments* – and indeed the researchers whose work is reprinted between its covers! – have their work cut out understanding the what, how and why of commonality and difference between the various real world cases interrogated. While the editor's four-part thematic division is useful in a heuristic sense, it inevitably offers few clues to the real points of connection and divergence between the published research. The irony, then, is this: the strength of geographical studies of nature's neoliberalisation – which I highlighted earlier – is also perhaps a current weakness. The focus on difference and path dependency has not yet, in my view, progressed to a careful sorting and sifting: that is, a comparative analysis that properly identifies 'signals in the noise' as well as variations in the 'noise' itself. Invocation of the term 'neoliberalism' in all of the studies printed here does not alter the fact that we currently lack ground-rules for comparing rigorously between empirical cases at a variety of scales. What we now need is a set of active engagements between researchers undertaking separate studies in separate times and places so that a spiral of learning can occur. Otherwise, it is left to readers of their research to do the work of comparative analysis, which can be a very tall order given the empirical and other differences between the cases interrogated. How and by what criteria do we identify the salient points of similarity and difference between real world instances of nature's neoliberalisation? As I said before, ideal-types like 'neoliberalism' do not provide a robust answer, but we cannot afford to get lost in the fine details of particular case studies either. So how do we navigate this particular Scylla and Charibdis?

Second, there are some pressing normative and policy issues. In a variety of ways, the chapters of *Neoliberal Environments* indicate a range of more-or-less severe problems attaching to nature's neoliberalisation. These problems in some way become the basis of implicit or explicit criticisms made by the chapter authors – all of whom, as I have said, are *critical* geographers. However, if their (and anyone else's) criticisms are to convince, then they must fulfil at least three requirements, in my view. First, these criticisms must be based on explicit and well-justified criteria. This is particularly important if those favouring neoliberalism are to have their beliefs challenged. Second, as already noted, they should be responsive to evidence, neither disregarding empirical findings that contradict the normative standpoints adopted nor becoming self-fulfilling prophecies because 'the facts' have been pre-determined to 'fit' these standpoints. Third, where criticisms of existing arrangements are made, a feasible alternative should be suggested (or at least hinted at) at the level of both principle and policy. The relationship between these three requirements is both direct and significant. The most philosophically robust value scheme can fail to persuade if there are no 'objective' reasons why this scheme is morally-ethically preferable to those prevailing in given real-world situations and no chance of it being practically achievable as a policy alternative. Without the second and third requirements, the first risks becoming dogma or pure utopianism.[4]

As the studies in this book reveal, to date critical geographers investigating nature's neoliberalisation have been much stronger on cognitive issues (description and explanation) than they have on practical policy and normative issues. Typically, policy alternatives remain implicit, while critical judgements take any number of tacit and more explicit forms. Critics of neoliberalism – be they in academic geography or any other walk of society – will be better served if robust policy alternatives are put forward, and if the maladies of neoliberalism are pin-pointed crisply and coherently. There is, of course, an ineradicable geography to policy alternatives and the evaluation of neoliberalisations too. A potentially vital contribution that critical geographers still can make is to show the blanket criticisms and blanket policy alternatives to neoliberalism are only plausible as heuristics or rallying calls. In the world of *realpolitik*, what is required is sensitive thinking about tailored alternatives to neoliberal policies based on the sorts of powerful criticisms of those policies that resonate in specific localities, regions and countries. This should not be the almost exclusive domain of policy works, think tanks and research NGOs.

Having touched upon policy issues, let me end with some comments on what is called 'public geography'. This book is one of many academic texts about neoliberalisation but one of just a few on the neoliberalisation of nature. The role such texts can play in informing wider public debate – as well as policy – remains uncertain. According to many commentators, research-led universities (at least here in the West) have become steadily divorced from the wider society in the post-war period: truly 'ivory towers'. This has some virtues, of course. For instance, without the relative autonomy that research universities have, the contributors in this book could not have undertaken their studies: studies which, after all, might not be possible if the universities were themselves to be successfully neoliberalised. However, there is a danger here that critical insights such as those expressed in this book remain *purely* academic: insights made by and for a coterie of what some disparagingly call 'tenured radicals'. There are a variety of ways round this, and far more examples already exist than most of us probably realise. What is required, it seems to me, is a way of building on these examples in order to contribute actively to organised attempts to make neoliberalism a thing of the past. Left-wing academics should not feel embarrassed or pompous when assuming a 'public' role – and after all, the 'public' in question might only be local or regional or sectional rather than the capital P national or international publics (always much harder to reach and influence if one is not already part of the tight world of journalism and broadcasting). To connect university research with the wider life of the societies that sustain it does not have to mean marketising, corporatising or 'dumbing-down' academic knowledge. There are ready audiences 'out there' for the sort of insights provided in a book like *Neoliberal Environments*.[5] What is now needed is a way of reaching them directly or by other means. Let us not forget that Hayek, back in the 1940s, was a little-known academic peddling what seemed for decades to be 'irrelevant' ideas that could have little purchase in the wider world. Critics of neoliberalism based in universities need to learn from this, if they have not already. There is always a way if the will exists in the first place. But, as ever, the way cannot be found in conditions of our own choosing. Herein lies the challenge for any opposition to neoliberalism, reformist or otherwise.

Notes

1 It would be a conceit to suggest that geographers alone do this, but it's nonetheless true that a core part of our *modus operandi* is to understand how commonality and difference together constitute time and space.

2 This said, where the research is currently lacking is the area of *biophysical* scale: that is, the complex links between nature's neoliberalisation of a resource in one place or region, and the links this process has with biophysical change elsewhere. Heynen and Perkins' Chapter 15 in this volume is among the few to broach this important question. Thus far, critical geographers have been more adept at dealing with the human dimensions of geographical scale.

3 This term, as I will explain, has a slightly different meaning to the editors' title choice *Neoliberal Environments*. The term 'environments' is nonetheless felicitous, even though I prefer the term 'ecology': it suggests not just that neoliberalism is remaking (and being remade by) the natural world. It also suggests that the *society* is being neoliberalised (albeit not always successful) so that the 'social environment' is conducive to further neoliberal projects in the future. Ryan Holifield's Chapter 16 in this book is an example of this attempted neoliberalisation of the social in conjunction with the biophysical.

4 This is not to discount the usefulness of thought experiments and forecasting. However, it is to place limits on the utility of critique where there is, practically speaking, no viable way to change that being criticised.

5 And some of them are currently 'in here': namely our students, who are tomorrow's decision-makers in government, business and civil society.

Conclusion

Unnatural consequences

Nik Heynen, James McCarthy,
Scott Prudham, and Paul Robbins

The cases reviewed in this volume bear witness to an incredibly imaginative and frankly disturbing set of experiments, from privatization of wild animals and banking of wetlands to deregulation of water quality. Do we have a name for these experiments? As people adapt and resist, are their responses and adaptations of a certain kind? And in naming these experiments and forms of resistance, and recognizing them as something more or less coherent, are we better poised to do anything about them?

To answer this, and because the discussion about neoliberalism has become so wide-ranging, we want to begin our conclusion by stressing our perspective that neoliberalism *is* capitalism, although a particular historical variant of capitalism. It is the most recent form of capitalism, one similar to, but also distinct from, classical liberalism and the *laissez-faire* liberalism discussed by Polanyi. We want to stress that, because it is capitalism, many of the features attributed to neoliberalism specifically are true of capitalism more generally. We also conclude by recognizing that capitalism has proved to be highly heterogeneous. Thus, Regulationists are right to try to identify characteristics of particular historically and geographically situated variants of capitalism; we see neoliberal capitalism over the past quarter century as one such variant, and we argue it is important to understand its particularities as well. The point here is that neoliberalism has sometimes been discussed, incorrectly in our view, as a single, monolithic and undifferentiated process that is somehow distinct from capitalism, rather than as a diverse and interlinked set of practices that reflect a heightened, evolved and more destructive form of capitalism.

That said, recent scholarship has rightly called for the implicit grounds of normative critiques to be made explicit and honed; the overwhelmingly critical research and scholarship on neoliberal natures has both an opportunity and an obligation to respond to this challenge by clarifying the linkages among its empirical, theoretical and normative projects. The research in this volume therefore comes at a critical moment. As research on the multiple geographies of neoliberalism referenced within this collection has proliferated, critical questions have emerged about how to proceed. The first question is whether "neoliberalism" is a valid analytical category, and if so, of what sort? Is it a Weberian ideal-type, an empirical generalization describing a conscious political project, a pejorative label academics apply to policies they dislike, or something else altogether?

Second, how do we recognize, "neoliberalism," amidst the complexities of real examples? That is, in this domain, what specific goals, measures, and practices does "neoliberalism" encompass, specifically with respect to environmental governance, and how do these disparate aspects all fit together? For instance, privatization and the creation of markets for "fictitious commodities" are clearly compatible, but how central are these to the neoliberalization of environmental governance relative to the rollback of command-and-control environmental regulations and increased roles in governance for non-state actors such as NGOs and communities?

Third, how might we best theorize such trends? For example, what analytical purchase is offered by Marxian political economy as opposed to governmentality or actor-network theory, or by institutional versus ethnographic research methods? On what grounds and by what criteria can and should we evaluate the neoliberalization of environmental governance?

We initially argued that collecting these works in one place adds value in several ways. After going through the process of producing this collection, we are convinced. First, both neoliberal discourses and critical debates about neoliberalism too often share the flaw of remaining abstract: too often, neither those who sing neoliberalism's praises nor those who parse its components do the work – or take the risks – of grounding their arguments in real examples. By contrast, this volume brings together specific case studies that span more than two decades of experience and evidence linking neoliberalism with concrete environmental changes, politics, and outcomes in diverse, international contexts. The chapters evaluate specific political ecologies and dynamics and the implications of particular reforms and enforcements, while collectively affording new contributors and readers the possibility of thinking comparatively across sectors and geographic contexts. Such specificity and comparative potential serve important analytical functions because they allow the authors and editors to craft stronger, more credible criticisms of common crises and outcomes, but also because they shed light on ways forward that are more just and more ecologically sound.

Precisely for this reason, this collection does political work. The contributors advance and refine both logical and substantive critiques of the neoliberalization of environmental governance both through deductive critiques of the internal contradictions of neoliberal thought and practice, and through case studies that document the effects of neoliberal reforms and the efficacy of resistance to them. These contributions, despite their pluarality, demonstrate that Margaret Thatcher's famous claim of "there is no alternative" to neoliberalism belies (1) the diversity of actual institutional and political projects mobilized under the rubric of neoliberal discourse; (2) the particular outcomes of neoliberal governance reforms; (3) the divergences between these outcomes and the promises by which neoliberalism is typically legitimated; and (4) the myriad and effective ways in which neoliberaliza- tions have been and continue to be resisted. In short, the work of the volume's many authors collectively offers the careful, empirically supported analysis and argumentation needed to challenge the abstractions that underpin pernicious orthodoxies. And it shows that like most orthodoxies, neoliberalism cannot be sustained.

The political utility of rhetorically involving neoliberalism also connects the work with countless activists and oppositional political movements around the world for whom the term carries clear meanings and normative associations. Such efforts to speak a common language are vital in reaching beyond the academy. Along the same lines, however, much as we may be critiquing capitalism when we critique neoliberalism, the latter seems to be a far more palatable language than a bald critique of "capitalism." Admit- tedly, a focus on critiquing the excesses of neoliberalism may point toward a more reformist politics than a direct critique of necessary features of capit- alism (i.e., you can keep wage labor, just let us have some sort of environ- mental and social safety net and some public spaces), but inasmuch as the latter would be a stunning political victory at the moment, we can live with it for the time being.

With respect to *environmental* governance, while neoliberal policies are of course varied and unpredictable in their outcomes (as Bakker has shown, for instance), we believe it is still reasonable to anticipate that they are more likely to produce negative than positive outcomes, insofar as their under- lying assumptions about markets and property are overwhelmingly mis- matched to the character and quality of biophysical systems, and their functions and flows. Beyond this, our concern about these negatives, these unnatural consequences, is further rooted in a precautionary evaluation of the incomplete information available to evaluate neoliberal capitalism's effect across spatial and temporal scales. Because environmental ramifica- tions of today's radical experiments will possibly be irreversible in so many cases (extinction of species, for instance), we feel it is imperative to take a stand now. Let history judge us all for our critical perspective if we are wrong and applaud us if we are not.

Instead of just assuming we are right, however, we must continue to undertake ecology and social science research in collaborative ways, with the same questions in mind, using robust methodologies to evaluate just how destructive these experiments are, how tractable they are to change, and how uneven they are in their effects. In this regard, it is important to maintain a view of neoliberalism in terms of hegemony and over- determination. Given the interdependencies and interrelatedness of agency, contingency, and complexity, how and why is it that people willingly choose the same items from the same tiny menu over and over? Why is there so much congruence in the menus and choices on offer around most of the world for some decades? Why do people make choices that maintain or exacerbate inequalities and undesirable status quos, and where do changes (that do sometimes occur!) come from? These are some of the ambitious questions that must be at the heart of our empirical work.

With this in mind, though, we suggest that work at the other end of the spectrum will be simultaneously essential, requiring us to roll our sleeves up and do environmental research to convincingly demonstrate what we suspect to be happening. In an era when climate scientists stand before Congress making undeniably political interventions against a status quo regime, we should not be afraid of engaging and cooperating with physical science and physical scientists. In a world where information, data, and evidence are increasingly available from diverse sources, we should not hesitate to consult secondary literatures and sources in the natural sciences. But more radically, at a time when the questions of social and physical sciences increasingly converge, we should not be afraid to retrain ourselves to interpret, communicate, and produce new forms of data outside the confines of our own disciplinary and sub-disciplinary training, and to train the next generation of scholars to be more wholly integrative. Political economic climatology, regulation hydrology, and subaltern wildlife ecology are *de facto* fields of research. We need to prepare ourselves to engage them.

We believe there is great analytical utility in naming and explaining neoliberalism. Enormous infringements on class privileges and powers, whether in the form of communism or strong welfare states, marked much of the prior century; neoliberalism is in large part about the effective elimination of such infringements and the political alternatives they represented, in order to create a space for the clawing back of the wealth, power, and privilege that the capitalist class was forced to cede. We see in neoliberalism, therefore, an international project to reclaim, reconstitute, or establish capitalist class privilege and power, dating from the late 1970s (see Harvey 2005). Clive Barnett (2005) is certainly correct that many elements of what we now characterize as neoliberalism are present as strands within the liberal tradition, but the question is which strands of liberalism are operationalized, in what ways, and to what effects in different periods. It seems to us madness, and a real failure, to fail to acknowledge that there has been an enormous global change over the past 25 years in the political context in which wealth is produced and distributed. How that works in particular contexts, including inside of people's heads, is of course always complicated, with any number of contingencies and internal contradictions at play, but to leave it at that is likely to miss the forest for the trees.

Finally, and despite our understanding of the natures around us as produced through politics and economy, we continue to hold to an explicitly normative environmental vision, one that holds hope for a cessation of potentially cataclysmic environmental changes in the world around us. The failed logic of neoliberalism and its ravenous craving for markets, commodities, and sites of accumulation across the planet, propels a loss of species that it has promised to defend, a destruction of ecosystems it has claimed to value, and a reduction in the quality of life that it professed to maintain. It is in need of replacement! We require utopian forms of environmental praxis to help us imagine alternative possibilities, emancipatory projects, and an end to social and environmental destruction at all scales. While communicating our skepticism toward the market enthusiasms so much a part of creeping neoliberal environmentalism, we require alliances with traditional members of the environmental community, and the green visions they carry and foment. If we are not willing to identify the possible range of environmental futures through our intellectual and political efforts, we are fated to produce an environmental outcome that mirrors the harsh, destructive logics that created it.

References

Barnett, C. (2005) "The Consolations of 'Neoliberalism,'" *Geoforum* 36: 7–12.
Harvey, D. (2005) *A Brief History of Neoliberalism*, Oxford: Oxford University Press.

Index

accumulation by dispossession *see* privatization

Africa 75, 78, 134, 194

agriculture 30, 130–31, 133, 231, 247; capital accumulation in 10; de/re-regulation of 126, 128, 171; domestic supports 15; ecological role of 132–35; European Model of 134; multifunctionality 128–29; subsidence 133, 135; trade 11, 15

agri-environmental 126–136

agro-industry 130–33, 171

Alabama 205

Aleutian Islands 69

Amazon Basin 194

American West *see* US West

Animal and Plant Health Inspection Service of the Department of Agriculture 30

anti-growth movement 141

Argentina 244, 277

Army Corps of Engineers (Corps), 115–16, 119–120

Asia 75, 134, 194

Beck, Ulrich 165, 174

Bering Sea 69

biodiversity 12, 114, 126, 128, 134–36, 157, 191, 222, 224

biotechnology 10

Blackwater Security 222

Blair, Tony 110

Bolivia 277

Brazil 16, 224–251, 269–71, 276–77

Brenner, Neil 3, 15, 94, 139, 153, 155, 193, 203

broken window theory 178,184

Cairns Group 128

California 29, 43–44

Canada 32, 39, 77, 79, 140–41, 164, 172, 177, 271–72

Canadian Methanex Corporation 43–44

cancer rates 44

capitalism 39, 51, 74, 91, 117, 121, 174, 187, 200, 218, 277, 289; alternative to 255–64; annihilation of space with time 193; contradictions of 26, 39, 47, 96, 193, 197, 199, 200; crisis 48, 154; destructive tendencies 15; "double movement" 274; green 17; ideology 1; interurban competition; nature 38, 45–6, 123, 190–95; neoliberal 287, 289

"capitalocentric", 256, 259, 260–61, 270–71

carbon dioxide 191, 219

carbon permits 11

Cardosa, Fernado Henrique 243–46

Castree, Noel 9, 102, 115

cattle industry 28

"Chicago Boys" 7, 275

Chile 7, 79, 275

cholera 96, 104, 143

Chronic Wasting Disease (CWD) 26, 31–36, 91, 96

civil society 4, 6, 10, 40, 92, 243, 251

class struggle 1, 36, 54

Clean Air Act 11

climate change 11, 105, 290

Clinton administration 202–205, 210–12, 218–19, 255

Cold War 1, 6, 8, 47, 274

Colorado 29

commodification 17, 56, 102, 154–55, 275; decommodification 16; definition of 103; of ecosystems 14–15, 114, 117,120, 122; of nature *see* nature; resistance to 107, 111, 156; of social